Planar Spiral Inductors, Planar Antennas and Embedded Planar Transformers

Amal Banerjee

Planar Spiral Inductors, Planar Antennas and Embedded Planar Transformers

SPICE-based Design and Performance Evaluation for Wireless Communications

 Springer

Amal Banerjee
Analog Electronics
Kolkata, India

ISBN 978-3-031-08780-6   ISBN 978-3-031-08778-3   (eBook)
https://doi.org/10.1007/978-3-031-08778-3

This Springer imprint is published by the registered company Springer Nature Switzerland AG
The registered company address is: Gewerbestrasse 11, 6330 Cham, Switzerland

This work is dedicated to:
My late father Sivadas Banerjee
My late mother Meera Banerjee
A dear friend, mentor, and guide
Dr. Andreas Gerstlauer

Contents

Chapter 1
Introduction and Problem Statement

1.1 Problem Statement

With wired|wireless telecommunication networks penetrating every corner of our planet, with ever-increasing carrier frequencies (e.g., 600 MegaHertz(MHz)-6 Giga-Hertz(GHz)-5G cellular phone system), telecommunication circuit designers have had to innovate|develop a novel integrated circuit, the miniaturized microwave integrated circuit or MMIC.

The key difference between any MMIC and all other analog|digital integrated circuits is as follows:

- **An inductor is an essential component of almost all RF|microwave electronic circuits, operating at 100 s of MHz–10s of GHz.** *So, any RF|microwave frequency range integrated circuit must include an inductor. An inductor (e.g., copper wire coil) stores energy in its magnetic field, analogous to a capacitor that stores energy in its electric field.*

   From physics, the inductance value of an inductor is directly proportional to its length, and to fit an inductor inside an MMIC, it must be shaped as a planar spiral. Theoretically, the spiral could be of any geometric shape. However, to analyze, design, and estimate the performance characteristics of a planar spiral inductor, both its physical and equivalent electrical circuit model must be known, as such a number of the equivalent electrical circuit model components depend on the physical model. The physical model is basically the answer to the following question:

- **What is the number of turns of a planar spiral inductor to satisfy a target total inductance value, within tight tolerances on trace width, trace separation, trace thickness, etc.?**

   The equivalent electrical circuit model components and their properties arise from the spiral shape of the inductor, *forcing current to flow in adjacent (nearest neighbor or otherwise) pair of planar spiral inductor segments in the same or opposite directions.* Currents in the same direction are termed as "even" and in the opposite direction "odd":

- For both even and odd modes, there is a parasitic capacitor between each inductor segment and the ground plane below it, separated by insulating dielectric layer.
- Parasitic capacitors between each antiparallel planar spiral inductor segment pair arise because the two opposite polarity segments act as a parallel plate capacitor.

Similarly, the final total inductance value includes the contribution from each planar spiral inductor segment (self-inductance) and the mutual inductance contribution from each possible (nearest neighbor and otherwise) pair of planar spiral parallel and antiparallel segments.

The parasitic capacitances, in combination with the even|odd mutual inductances, forces parallel|antiparallel current flow planar spiral inductor segment pairs to act as even|odd coupled transmission lines. So, transmission line theory concepts and results can be exploited to create the equivalent circuit model of the planar spiral inductor. In network theory terms, a planar spiral inductor is a two-port network. Capacitive and inductive coupling between parallel microsrips, when the current is flowing through them in opposite directions is *crosstalk.*

In addition to the actual planar spiral inductor design and its performance characteristic evaluation, the RF|microwave circuit designer must also focus on how to fabricate it, given various manufacturing constraints. Given that an integrated circuit consists of vertically stacked layers of transistors, signal lines, ground planes, and power lines, some key constraints are as follows:

- To create the planar spiral inductor pattern on the integrated circuit layer, its corresponding mask has to be created first. This mask must ensure that the separation between any two nearest neighbor inductor segments is constant, within tight predefined tolerances.
- The planar spiral inductor has to be connected to other circuit components, often located in layers above and below it.
- The planar spiral inductor must be shaped so that it makes optimum use of available integrated circuit layer floor area.

Transformers consisting of planar spiral inductors and embedded inside an integrated circuit are a topic of intense research nowadays. They are a direct application of planar spiral inductors. There are two possible types of these transformers. The coplanar embedded transformer has both its primary and secondary inductors such that each segment of the secondary planar spiral inductor has as its nearest adjacent neighbor a segment of the primary planar spiral inductor, and vice versa. The vertically stacked embedded transformer has its primary and secondary planar spiral inductors in vertically adjacent planes. This type of planar embedded planar transformer can provide dual polarity output at its secondary, by placing the primary planar spiral inductor in a plane in between the two secondary planar spiral inductors.

The equivalent electrical circuit model for this transformer is the same as that of coupled transmission lines, and both mutual induction and capacitive coupling between segments of coplanar or vertically stacked planar spiral inductors transfer electromagnetic energy from the primary to the secondary inductor, and the equivalent electrical circuit is similar to that of two transmission lines with *crosstalk*. This is straightforward to simulate in SPICE [1–6].

The manufacturing constraints are more stringent than that of a standalone planar spiral inductor. In addition to the three constraints mentioned earlier, two new ones apply.

- For a coplanar embedded planar transformer, the primary|secondary inductor segment in between two secondary|primary must be positioned so that capacitive coupling between nearest adjacent segments is the same (within applicable tolerances).
- For vertically stacked embedded planar spiral transformer, the trace widths of the primary and secondary planar spiral inductors must be the same (within applicable tolerances), and the planar inductors must be aligned along the common vertical axis (within applicable tolerances).

Planar antennas are called "patch" antennas as they look like patch of metal. These are found in all cell phones(total three—Bluetooth, Wi-Fi, and primary cellular) and daily use SOHO (small office/home office) routers, switches, etc. These devices use either the planar patch antenna, the planar inverted F antenna (PIFA), the planar loop antenna, or the planar dipole antenna.

Like the planar spiral inductor|embedded planar transformer, the analysis, design, and performance characteristic estimation of planar antennas are based on microstrip transmission line theory. The key difference is that a planar antenna must be ***impedance matched*** to the receiver|transmitter, since the signal receiver|transmitter and free space have different impedances. So the transmitter output impedance must be matched to the antenna input impedance to minimize signal reflection and signal energy loss. The intermediate transmission line (between signal receiver and transmitter) and planar antenna can be connected to the antenna edge|a distance in (inset) connection. Alternatively, it can be connected with an inset metal pin.

The restrictions of fitting the antenna inside an integrated circuit or connecting it to other circuit components in layers above or below it are absent. So researchers have suggested various shapes for these antennas—circular strip, disc, semicircle, circular segment, triangle, and the familiar rectangle and square.

Numerous researcher(s) [7–53] have analyzed the underlying physics of planar spiral inductors, and most of their (**barring** [20, 21]) analysis, conclusions, and results have no useful value.

- Each researcher|research team has its unique approach|method|scheme, often very mathematical to analyze, design, and evaluate the performance characteristics of planar spiral inductors. Unfamiliarity with the presented inductor analysis scheme makes it very difficult to extract the essential design equations|formulas or the sequence of calculation steps used. Also, unstated simplifying assumptions (used by the article|paper authors) dilute the problem to be solved. So experimentally obtained results do not match formula-based calculation results, within n applicable tolerances.

- Often published equations and formulas contain constants obtained by numerical nonlinear curve fitting of experimental data. These constant values can change depending on initial values supplied to the nonlinear curve fitting code, and often need adjustment for convergence. Such formulas are relevant only in that context alone.
- Often in-house or proprietary [54, 55] CAD (computer-aided design) tools are used to analyze|design a planar spiral inductor, so that the results are meaningful in that context of the CAD tool only. The published results are confusing, as no details about the calculation steps can be provided (algorithms used by the CAD tool are proprietary). A few CAD tools include basic information about the underlying theoretical basis for their own CAD tool, to entice new customers. Unfortunately, these proprietary CAD tools [54, 55] are expensive (to install| configure|maintain) and have steep learning curves. Interestingly, the core electrical|electronic circuit analysis|simulation engine of these CAD tools is SPICE (Simulation Program with Integrated Circuit Emphasis) [1–6].
- Researchers involved in planar antenna analysis|design have spent effort and time deriving new designs as circular ring, disc, circular segment, triangle, vertically stacked triangles, etc., which are not used in real-world electronic devices, where the prime concern is how to optimize available printed circuit board floor area, for which only the rectangle and square are best.
- Tried and used electrical|electronic circuit simulation and performance characteristic evaluation software tool is available as both open-source and proprietary versions.
- To circumvent the planar spiral inductor|embedded transformer|antenna manufacturability constraints, the proposed solution scheme adapts the semiconductor industry standard that all planar spiral inductors are rectangular or square shaped.

1.2   The Solution

This book demonstrates a detailed, unified, and bottom-up scheme to analyze, design, and estimate key performance characteristics of planar spiral inductors, embedded (on-chip) planar spiral inductor transformers, and planar antennas.

- The analysis|design|performance characteristic evaluation scheme has been designed to answer the key question: how many series connected planar inductors are needed to achieve a predefined target value?
- This scheme is based on Maxwell's laws of classical electrodynamics (electromagnetic wave transmission through conductors in dielectric substrates) and transmission line theory (microstrip transmission lines).

- The bottom-up method is embodied in the fact that the analysis scheme starts with a single planar rectangular or circular loop inductor and using that information to examine the case of a set of series connected rectangular inductors that form a planar spiral inductor [20, 21]. This scheme also examines in detail modern current sheet [24] approach to analyze planar spiral inductors.
- The key information gleaned from planar spiral inductor analysis is applied to examine coplanar or vertically stacked planar spiral inductors that form an on-chip embedded planar spiral inductor transformer.
- For both standalone planar spiral inductors and embedded planar spiral inductor transformers, the analysis includes unavoidable parasitic capacitances and inductances that arise because these planar spiral inductors operate at 100 s of MHz or 10s of GHz.
- The analysis scheme includes planar antennas that operate at resonance, impedance matched to maximize signal power reception or transmission. This is unlike the normal mode for planar spiral inductors and embedded planar spiral inductor transformers.
- As the design calculations are complicated, computer programs are essential to evaluate the component values for the equivalent electrical circuit for a planar spiral inductor. These computed component values are used for the corresponding SPICE [1–5] netlist. SPICE [1–5] is the gold standard electrical|electronic circuit performance tool and is available as open-source and proprietary versions. **Therefore, to avoid the use of expensive CAD (computer-aided design) [6, 54, 55] tools, a set of ANSI C computer language [56] executables (for both the popular Linux and Windows operating systems) is supplied to perform the calculations for evaluation of the component values for the equivalent electrical circuit for a planar spiral inductor and most importantly format the computed values in a SPICE [1–5] text input format netlist. This editable netlist can be used with any (open-source|proprietary) available SPICE [1–5] simulator.**

The details of this novel scheme are elaborated on in the subsequent chapters.

References

1. http://ngspice.sourceforge.net/docs/ngspice-manual.pdf
2. https://ecee.colorado.edu/~mathys/ecen1400/pdf/scad3.pdf
3. https://www.seas.upenn.edu/~jan/spice/PSpice_UserguideOrCAD.pdf
4. https://cseweb.ucsd.edu/classes/wi10/cse241a/assign/hspice_sa.pdf
5. https://ece.northeastern.edu/courses/eece7353/2019sp/hspice_sa.pdf
6. https://www.emisoftware.com/calculator/microstrip/
7. Long, J. R., & Copeland, M. A. (1997). The modeling, characterization and design of monolithic inductors for silicon RF ICs. *IEEE Journal of Solid-State Circuits, 32*, 357–369.
8. Niknejad, A. M., & Meyer, R. G. (1998). Analysis, design, and optimization of spiral inductors and transformers for Si RF ICs. *IEEE Journal of Solid-State Circuits, 33*, 1470–1481.

9. Reyes, A. C., El-Ghazaly, S. M., Dorn, S. J., Dydyk, M., Schroder, D. K., & Patterson, H. (1995). Coplanar waveguides and microwave inductors on silicon substrates. *IEEE Transactions on Microwave Theory and Technology, 43*, 2016–2022.
10. Ashby, K. B., Koullias, I. C., Finley, W. C., Bastek, J. J., & Moinian, S. (1996). High Q inductors for wireless applications in a complementary silicon bipolar process. *IEEE Journal of Solid-State Circuits, 31*, 4–9.
11. Lu, L. H., Ponchak, G. E., Bhattacharya, P., & Katehi, L. (2000). High-Q X-band and K'-band micromachined spiral inductors for use in si-based integrated circuits. *Proceedings of Silicon Monolithic Integrated Circuits RF Systems*, 108–112.
12. Bahl, I. J. (1999). Improved quality factor spiral inductor on gaas substrates. *IEEE Microwave Guided Wave Letters, 9*, 398–400.
13. Ribas, R. P., Lescot, J., Leclercq, J. L., Bernnouri, N., Karam, J. M., & Courtois, B. (1998). Micromachined planar spiral inductor in standard GaAs HEMT MMIC technology. *IEEE Electron Device Letters, 19*, 285–287.
14. Takenaka, H., & Ueda, D. (1996). 0.15μm T-shaped gate fabrication for GaAs MODFET using phase shift lithography. *IEEE Transactions on Electron Devices, 43*, 238–244.
15. Chiou, M. H., & Hsu, K. Y. J. (2006). A new wideband modeling technique for spiral inductors. *IET Microwave, Antennas, and Propagation, 151*, 115–120.
16. Lu, H.-C., Chan, T. B., Chen, C. C. P., & Liu, C. (2010). M, spiral inductor synthesis and optimization with measurement. *IEEE Transactions on Advanced Packaging, 33*.
17. Talwalkar, N. A., Yue, C. P., & Wong, S. S. (2005). Analysis and synthesis of on-chip spiral inductors. *IEEE Transactions on Electron Devices, 52*, 176–182.
18. Mukherjee, S., Mutnury, S., Dalmia, S., & Swaminathan, M. (2005). Layout-level synthesis of RF inductors and filters in LCP substrate for Wi-Fi applications. *IEEE Transactions on Microwave Theory and Technology, 53*, 2196–2210.
19. Kulkarni, J. P., Augustine, C., Jung, C., & Roy, K. (2010). Nano spiral inductors for low-power digital spintronic circuits. *IEEE Transactions on Magnetics, 46*, 1898–1901.
20. Greenhouse, H. M. (1974). Design of planar rectangular microelectronic inductors. *IEEE Transactions on Parts, Hybrids and Packaging, 10*, 101–109.
21. https://books.google.co.in/books?hl=en&lr=&id=K3KHi9lIltsC&oi=fnd&pg=PR13& dq=grover+inductance+calculations&ots=dPY1K2rxOd&sig=ZLMajbfyFc0P4 EBh5wukPJZBa8w#v=onepage&q=grover%20inductance%20calculations&f=false
22. Jenei, S., Nauwelaers, B. K. J. C., & Decoutere, S. (2002). Physics-based closed-form inductance expression for compact modeling of integrated spiral inductors. *IEEE Journal of Solid-State Circuits, 37*, 77–80.
23. Asgaran, S. (2002). New accurate physics-based closed-form expressions for compact modeling and design of on-chip spiral inductors. *Proceedings of the 14th International Conference on Microelectronics*, 247–250.
24. Mohan, S. S., Hershenson, M. M., Boyd, S. P., & Lee, T. H. (1999). Simple accurate expressions for planar spiral inductance. *IEEE Journal of Solid-State Circuits, 34*, 1419–1424.
25. Chen, C. C., Huang, J. K., & Cheng, Y. T. (2005). A closed-form integral model of spiral inductor using the Kramers-Kronig relations. *IEEE Microwave and Wireless Components Letters, 15*.
26. Sieiro, J., Lopez-Villegas, J. M., Cabanillas, J., Osorio, J. A., & Samitier, J. (2002). A physical frequency-dependent compact model for RF integrated inductors. *IEEE Transactions on Microwave Theory and Technology, 50*, 384–392.
27. Sun, H., Liu, Z., Zhao, J., Wang, L., & Zhu, J. (2008). The enhancement of Q-factor of planar spiral inductor with low-temperature annealing. *IEEE Transactions on Electron Devices, 55*, 931–936.
28. Tsai, H. S., Lin, L., Frye, R. C., Tai, K. L., Lau, M. Y., Kossives, D., Hrycenko, F., & Chen, Y. K. (1997). Investigation of current crowding effect on spiral inductors. *IEEE MTT-S Symposium on Technologies to Wireless Applications*, 139–142.

29. Bushyager, N., Davis, M., Dalton, E., Laskar, J., & Tentzeris, M. (2002). Q-factor and optimization of multilayer inductors for RF packaging microsystems using time domain techniques. *Electronic Components and Technology Conference*, 1718–1721.
30. Eroglu, A., & Lee, J. K. (2008). The complete design of microstrip directional couplers using the synthesis technique. *IEEE Transactions on Instrumentation and Measurement, 12*, 2756–2761.
31. Costa, E. M. M. (2009). Parasitic capacitances on planar coil. *Journal of Electromagnetic Waves and Applications, 23*(17–18), 2339–2350.
32. Nguyen, N. M., & Meyer, R. G. (1990). Si IC-compatible inductors and LC passive filter. *IEEE Journal of Solid-State Circuits, 27*(10), 1028–1031.
33. Zu, L., Lu, Y., Frye, R. C., Law, Y., Chen, S., Kossiva, D., Lin, J., & Tai, K. L. (1996). High Q-factor inductors integrated on MCM Si substrates. *IEEE Transactions on Components. Packaging and Manufacturing Technology, Part B: Advanced Packaging.*
34. Burghartz, J. N., Soyuer, M., & Jenkins, K. (1996). Microwave inductors and capacitors in standard multilevel interconnect silicon technology. *IEEE Transactions on Microwave Theory and Technology, 44*(1), 100–103.
35. Merrill, R. B., Lee, T. W., You, H., Rasmussen, R., & Moberly, L. A. (1995). Optimization of high Q integrated inductors for multi-level metal CMOS. *IEDM*, 38.7.1–38.7.3.
36. Chang, J. Y. C., & Abidi, A. A. (1993). Large suspended inductors on silicon and their use in a 2 μm CMOS RF amplifier. *IEEE Electron Device Letters, 14*(5), 246–248.
37. Craninckx, J., & Steyaert, M. (1997). A 1.8-GHz low-phase-noise CMOS VCO using optimized hollow spiral inductors, *IEEE Journal of Solid-State Circuits, 32*(5), 736–745.
38. Lovelace, D., & Camilleri, N. (1994). Silicon MMIC inductor modeling for high volume, low cost applications. *Microwave Journal*, 60–71.
39. Kamon, M., Tsulk, M. J., & White, J. K. (1994). FASTHENRY a multipole accelerated 3-D inductance extraction program. *IEEE Transactions on Microwave Theory and Technology, 42*(9), 1750–1757.
40. Pettenpaul, E., Kapusta, H., .Weisgerber, A., Mampe, H., Luginsland, J., & Wolff, I. (1988). CAD models of lumped elements on GaAs up to 18 GHz, *IEEE Transactions of Microwave Theory and Technology, 36*(2), 294–304.
41. Howard, G. E., Yang, J. J., & Chow, Y. L. (1992). A multipipe model of general strip transmission lines for rapid convergence of integral equation singularities. *IEEE Transactions on Microwave Theory Technology, 40*(4), 628–636.
42. Gharpurey, R. (1995). *Modeling and analysis of substrate coupling in integrated circuits.* Doctoral thesis, University of California, Berkeley.
43. Stetzler, T., Post, I., Havens, J., & Koyama, M. (1995). A 2.7V to 4.5V single-chip GSM transceiver RF integrated circuit IEEE. *International Solid-State Circuits Conference*, 150–151.
44. Kim, B. K., Ko, B. K., Lee, K., Jeong, J. W., Lee, K.-S., & Kim, S. C. (1995). Monolithic planar RF inductor and waveguide structures on silicon with performance comparable to those in GaAs MMIC. *IEDM*, 29.4.1–4.4.
45. Krafesik, D., & Dawson, D. (1986). A closed-form expression for representing the distributed nature of the spiral inductor. *Proceedings of the IEEE-MTT Monolithic Circuits Symposium*, 87–91.
46. Kuhn, W. B., Elshabini-Riad, A., & Stephenson, F. W. (1995). Centre-tapped spiral inductors for monolithic bandpass filters. *Electronics Letters, 31*(8), 625–626.
47. Kurup, H. B., Dinesh, S., Ramesh, M., & Rodrigues, M. (2020). Low profile dual-frequency shorted patch antenna. *International Journal of recent Technology and Engineering ISSN: 2277–3878, 8*(5).
48. Mishra, A., Singh, P., Yadav, N. P., Ansari, J. A., & Viswakarma, B. R. (2009). Compact shorted microstrip patch antenna for dual band operation. *Progress In Electromagnetics Research C, 9*, 171–182.

49. Ansari, J. A., Singh, P., Yadav, N. P., & Viswakarma, B. R. (2009). Analysis of shorting pin loaded half disk patch antenna for wideband operation. *Progress in Electromagnetics Research C, 6*, 179–192.
50. Kathiravan, K., & Bhattacharyya, A. K. (1989). Analysis of triangular patch antennas. *Electromagnetics, 9*(4), 427–438. https://doi.org/10.1080/02726348908915248
51. Malik, J., & Kartikeyan, M. (2011). A stacked equilateral triangular patch antenna with Sierpinski gasket fractal for WLAN applications. *Progress In Electromagnetics Research Letters, 22*, 71–81. https://doi.org/10.2528/PIERL10122304. http://www.jpier.org/PIERL/pier.php?paper=10122304
52. Tripathi, A. K., Bhatt, P. K., & Pandey, A. K. (2012). A comparative study of rectangular and triangular patch antenna using HFSS and CADFEKO. *International Journal of Computer Science and Information Technologies, 3*(6), 5356–5358.
53. Balanis, C. A. Antenna Theory Analysis and Design Fourth Edition Copyright 2016 by John Wiley & Sons, Inc Library of Congress Cataloging-in-Publication Data:ISBN 978–1–118-642060-1 (cloth) 1. Antennas (Electronics) I. Title.TK7871.6.B354 2016 621.382.
54. https://www.awr.com/awr-software/products/microwave-office
55. https://www.keysight.com/in/en/products/software/pathwave-design-software/pathwave-advanced-design-system.html
56. The all time classic C computer language book by its creators is available at: https://www.amazon.in/Programming-Language-Prentice-Hall-Software/dp/0131103628

Chapter 2
Fundamental Physics of Planar Inductors, Embedded Planar Transformers, and Planar (Patch) Antennas

2.1 Maxwell's Laws of Electrodynamics for Conductors

At radio frequency (RF) and microwave frequency ranges (100 s of MHz–10s of GHz), propagation of electromagnetic waves in both conductors and dielectrics (insulators) is governed by Maxwell's equations for electrodynamics, which in their most general form are [42]

$$\nabla \vec{D} = \rho \quad \nabla \vec{B} = 0 \quad \nabla \vec{x} \vec{B} = \vec{J} + \frac{\partial \vec{D}}{\partial t} \quad \nabla \vec{x} E = \frac{-\partial \vec{H}}{\partial t} - \vec{M}$$

where each of the vector quantities is $\vec{E}, \vec{H}$ electric and magnetic field intensities, $\vec{D}, \vec{B}$ electric and magnetic field densities $\vec{J}, \vec{M}$ electric and fictitious magnetic current densities, and ρ electric charge density, respectively. The units for each of these quantities are MKS.

The properties of electromagnetic waves and Maxwell's equations in free space are not very interesting and will not be examined in any detail. Interesting phenomena occur as the electromagnetic wave propagates through a conductor or dielectric.

For a dielectric, an applied electromagnetic field polarizes the material, creating a dipole polarization moment, that adds to the total displacement flux $\vec{D} = \epsilon_0 \vec{E} + \vec{P}$ where ϵ_0 is the permittivity of free space. In a linear medium, the polarization vector is $\vec{P} = \epsilon_0 \chi_e \vec{E}$ where the electric susceptibility is complex, and then the displacement flux is re-written as $\vec{D} = \epsilon_0 (1 + \chi_9) \vec{E}$. The imaginary part of the complex permittivity is responsible for the energy loss as the electromagnetic wave propagates through the dielectric medium, caused by the heating up of the material molecules as the wave propagates. The dielectric material molecules absorb energy from the electromagnetic waves and then release it—heating up the bulk. The dielectric energy loss is equivalent to conductor loss—for a conductor, the current density is

© The Author(s), under exclusive license to Springer Nature Switzerland AG 2023
A. Banerjee, *Planar Spiral Inductors, Planar Antennas and Embedded Planar Transformers*, https://doi.org/10.1007/978-3-031-08778-3_2

$\vec{J} = \sigma\vec{E}$ (Ohm's law in electromagnetic wave theory terms) where σ is the material conductivity. Then Maxwell's law re-written with these substitutions becomes

$$\nabla x \vec{H} = \left[\sigma + \frac{\partial(1 + \chi_0)}{\partial t}\right]\vec{E}$$

which can be simplified to

$$\nabla x \vec{H} = j\omega\left[\epsilon_{REAL} - j\epsilon_{IMG} - \frac{j\sigma}{\omega}\right]\vec{E},$$

and the loss tangent (ratio of real to imaginary parts of the displacement current) can be defined as $\tan\delta = \frac{j\epsilon_{MGG} + \sigma}{\omega\epsilon_{REAL}}$.

The real part of the dielectric constant and the loss tangent are used to classify microwave materials. The dielectric constant can be re-written in terms of the loss tangent as $\epsilon = \epsilon_{REAL} - j\omega\tan\delta$.

So far, it has been assumed that the polarization is a vector, implying a uniform, isotropic material. In case of non-isotropic material, the expressions involving $\vec{D}, \vec{E}, \vec{P}$ include a tensor of rank 2.

For a magnetic material, the relation between magnetic flux density and magnetic field intensity includes a magnetic polarization term:

$$\vec{B} = \mu_0\left(\vec{H} + \vec{P}_{magnetic}\right).$$

For a linear magnetic material,

$$\vec{P}_{nagnetic} = \chi_{magnetic}\vec{H},$$

so that

$$\vec{B} = \mu_0\left(\vec{H} + \vec{P}_{magnetic}\right) = \mu_0(1 + \chi_{magnetic})\vec{H} = \mu\vec{H}.$$

The magnetic permeability is complex, so that

$$\mu = \mu_9\left(1 + \chi_{magnetic}\right) = \mu_{REAL} - j\mu_{IMG}.$$

For a non-isotropic magnetic material, the magnetic material is a tensor of rank 2. For a linear medium, none of the electric permittivity or magnetic permeability depend on the electric or magnetic field intensities, and Maxwell's equations can be re-written as

$$\nabla \vec{B} = 0 \nabla \vec{D} = \mu \nabla x \vec{H} = j\omega\epsilon \vec{E} + \vec{J} \nabla x \vec{E} = -j\omega\mu \vec{H} - \vec{M}.$$

At the interface between two dielectric media, the normal components of $\vec{B}, \vec{D}$ are continuous, and the tangential components of $\vec{E}, \vec{H}$ are continuous. These conditions can be summarized as

$$\hat{n}_1 \vec{B}_1 = \hat{n}_2 \vec{B}_2 \hat{n}_1 \vec{D}_1 = \hat{n}_2 \vec{D}_2 \hat{n}_1 \vec{E}_1 = \hat{n}_2 E_2 \hat{n}_1 \vec{H}_1 = \hat{n}_2 H_2$$

Similarly, for the interface between two ideal conductors, all field components must be zero inside the conducting region, and the corresponding boundary conditions are

$$\hat{n} \vec{B} = 0 \hat{n} \vec{D} = \rho_{SURFACE} \hat{n} \times \vec{E} = 0 \hat{n} \times \vec{H} = \vec{J}_{SURFACE}$$

This is the **electric wall** boundary condition.

A very interesting boundary condition is the **magnetic wall**, which is a result of the tangential component of the magnetic field intensity being zero at the interface. $\hat{n} \times \vec{H} = 0$ such as on corrugated surfaces or at planar transmission lines—and is analogous to the current and voltage at the extremity of an open circuited transmission line. So, the magnetic wall can be summarized as

$$\hat{n} \vec{B} = 0 \hat{n} \vec{D} = 0 \hat{n} \times \vec{E} = -\vec{M} \hat{n} \times \vec{H} = 0$$

where $\vec{M}$ is the fictitious magnetic current, introduced for convenience, with the unit normal pointing out of the magnetic wall region.

In a linear, isotropic, homogeneous source-free region, the electric and magnetic field intensities are related as

$$\nabla \times \vec{E} = -j\omega\mu \vec{H} \nabla \times \vec{H} = j\omega\epsilon \vec{E}$$

Substituting one equation in the other, and using the useful expression,

$$\nabla \times \nabla \times \vec{E} = \nabla \nabla \vec{E} - \nabla^2 \vec{E}$$

gives the simple plane wave equation
$\nabla^2 \vec{E} + \omega^2 \mu\epsilon \vec{E} = 0$. A similar equation for the magnetic field intensity is

$$\nabla^2 \vec{H} + \omega^2 \mu\epsilon \vec{H} = 0$$

In a lossless medium, both the dielectric constant and magnetic permeability are real, and the wave number k is real. The solution to the above equation can be found by considering an electric field in the, e.g., x direction and invariant in the y, z directions, which gives

$$\frac{\partial^2 E_X}{\partial X^2} + k^2 E_X = 0$$

The general solution to this is $E_X(z) = E_{PLUS}e^{-jkz}E_{MINUS}e^{jkz}$.
where the "PLUS|MINUS" refer to propagation in the +ve|-ve z directions. In general, the solution is written as

$$E_X(z,t) = E_{PLUS}\cos(\omega t - kz) + E_{MINUS}\cos(\omega t - kz)$$

The phase velocity and wavelength of this propagating wave are

$$v_p = \frac{\omega}{k} = \frac{1}{\sqrt{\mu\epsilon}} \quad \text{and} \quad \lambda = \frac{2\pi}{k}$$

The corresponding magnetic field component can be expressed in terms of the electric field component, keeping in mind that it is orthogonal to the electric field. Once either the electric or magnetic field d is known, the other can be dediced in a straightforward way:

$$H_y = \frac{1}{\eta}\left(E_{PLUS}e^{-jkz} - E_{MINUS}e^{jkz}\right)$$

where the wave impedance is defined as $\eta = \frac{\omega\mu}{k}$.

Real-world materials are all non-isotropic or inhomogeneous or nonlinear, or a combination of these, and so is conductive; the coupled Maxwell's equations can be re-written as

$$\nabla \times \vec{E} = -j\omega\mu\vec{H}\nabla \times \vec{H} = \vec{J} + j\omega\epsilon\vec{E},$$

so that using similar substitutions as before,

$$\nabla^2\vec{E} + \mu\epsilon\left(1 - \frac{J\sigma}{\omega\epsilon}\right)\vec{E} = 0.$$

The extra damping term, which contributes to wave attenuation, is clear.
The complex propagation constant can be defined as

$$\gamma = \alpha + j\beta = j\omega\sqrt{\mu\epsilon}\left(1 - \frac{j\sigma}{\omega\epsilon}\right).$$

As before, if the electromagnetic wave has an x component only, the wave equation is reduced to

$$\frac{\partial^2 E_x}{\partial x^2} - \gamma^2 E_x = 0,$$

and the general solution for this equation is

$$E_x(z) = E_{PLUS}e^{-\gamma z} + E_{MINUS}e^{\gamma z} \quad \text{where} \quad e^{-\gamma z} = e^{-\alpha z + j\beta z},$$

which in the real time domain is

$$e^{-\gamma z} = e^{i\alpha z}\cos(\omega t - \beta z) \quad \text{and} \quad \gamma = j\omega\sqrt{\mu\epsilon} = jk = j\omega\sqrt{\mu\epsilon_{REAL}(1 - j\tan\delta)}.$$

The corresponding magnetic field component is

$$H_y = \frac{1}{\eta}(E_x e^{-\gamma z} - E_x e^{\gamma z}) \quad \text{where} \quad \eta = \frac{j\omega\mu}{\gamma}.$$

the wave impedance is complex, and reduces to the lossless form in a lossless medium.

For a perfect conductor, *the conductive current is much larger than the displacement current*. This means that the imaginary part of the dielectric constant is much greater than the real part, i.e.,

$$\epsilon_{IMG} \gg \epsilon_{REAL}.$$

This means that

$$\gamma = \alpha + j\beta = j\omega\sqrt{\mu\epsilon}\sqrt{\frac{\sigma}{j\omega\epsilon}} = (1 + j)\sqrt{\frac{\omega\sigma\mu}{2}}.$$

The term $\frac{1}{\alpha} = \sqrt{\frac{\omega\mu\sigma}{2}} = \delta$ is called the skin depth and physically indicates the depth inside the conductor at which the electric field intensity decreases to $\frac{1}{e}$ of its value at the surface of the conductor. *The physical implication of this is that only a very thin conductor can be used in circuits for RF and microwave frequencies.*

In the most general case, the wave equation in free space can be described as

$$\nabla^2\vec{E} + K_0^2\vec{E} = \frac{\partial^2 E_x}{\partial x^2} + \frac{\partial^2 E_y}{\partial y^2} + \frac{\partial^2 E_z}{\partial z^2} + k_0^2\vec{E} = 0$$

whose general solution is of the form

$$E(x, y, z) = E(x)E(y)E(z).$$

The solution is

$$E(x, t, z) = f(x)g(y)h(x)$$

and separation of variables technique is used to solve the above partial differential equation. This generates three differential equations as

$$\frac{d^2 f}{dx^2} + k_x^2 f = 0 \quad \frac{d^2 g}{dy^2} + k_y^2 g = 0 \quad \text{and} \quad \frac{d^2 h}{dz^2} + k_z^2 h = 0.$$

Each of these has simple harmonic oscillator solutions, and $k_x^2 + K_y^2 + k_z^2 = k_0^2$. The general solution for the electric field component in the x direction is

$$E_x(x, y, z) = E_{x0} e^{-j\left(k_x x + k_y y + k_z z\right)},$$

and similar solutions exist for the y, z directions. Using $\gamma = X\widehat{X} + Y\widehat{Y} + Z\widehat{Z}$ the solutions to the partial differential equation, using the separation of variables, are

$$E_x(x, y, z) = E_{x0} e^{-j\vec{k}\cdot\vec{r}}, \quad E_y(x, y, z) = E_{y0} e^{-j\vec{k}\cdot\vec{r}}, \quad \text{and}$$

$$E_z(x, y, z) = E_{z0} e^{-j\vec{k}\cdot\vec{r}}.$$

To satisfy Maxwell's equations, the three components must be orthogonal to each other, and so the divergence condition implies that

$$\vec{\nabla} \cdot \vec{E} = \frac{\partial E_x}{\partial x} + \frac{\partial E_y}{\partial y} + \frac{\partial E_z}{\partial z} = 0$$

In addition, using $\vec{E_0} = E_{x0}\widehat{x} + E_{y0}\widehat{y} + E_{z0}\widehat{z}$ the divergence condition gives

$$\vec{\nabla} \cdot \vec{E} = -j\vec{k} \cdot \vec{E} e^{-jk_z z} = 0 \quad \text{so that} \quad \vec{k} \cdot \vec{E} = 0.$$

Reapplying the divergence condition gives

$$\vec{\nabla} \times \vec{E} = -j\mu_0 \omega \vec{H}$$

which then gives the magnetic field intensity(which after some manipulation) which is

$$\vec{H} = \frac{1}{\eta_0} \widehat{n} \times \vec{E} \quad \text{where} \quad \eta_0 = \sqrt{\frac{\mu_{0_i}}{\epsilon_0}}.$$

The electric field can be written as

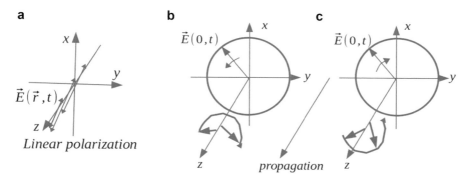

Fig. 2.1 (**a**) Linearly polarized electromagnetic wave. Double-ended arrows indicate electric field vector. (**b**, **c**) Right and left circularly polarized electromagnetic wave

$$\vec{E}(x, y, z, t) = \vec{E}_0 \cos\left(\vec{k} \cdot \vec{r} - \omega t\right)$$

A large sheet of flowing charge serves as a source of electromagnetic waves. If the surface current density $(\vec{J}s = J_x\hat{x})$ exists in the $z = 0$ plane (equivalently x, y plane) in free space, and since the source does not vary in the x, y directions, the electric fields are also invariant in the same directions. Denoting $\vec{E}_1, \vec{H}_1$ as the electric fields for $z < 0$, and $\vec{E}_2, \vec{H}_2$ the corresponding fields for $z > 0$, the following two conditions hold:

$$\hat{n} \times \left(\vec{E}_2 - \vec{E}_{1_i}\right) = \hat{z} \times \left(\vec{E}_2 - \vec{E}_1\right) = 0 \text{ and}$$

$$\hat{n} \times \left(\vec{H}_2 - \vec{H}_{1_i}\right) = \hat{z} \times \left(\vec{H}_2 - \vec{H}_1\right) = J_X\hat{X}.$$

Then the electric and magnetic fields for $z < 0$ and $z > 0$ are respectively

$$\vec{E}_1 = \hat{x}\eta_0 A e^{jk_0 z}, \vec{H}_1 = -\hat{y} A e^{jk_0 z} \text{ and}$$

$$\vec{E}_2 = \hat{x}\eta_0 B e^{-jk_0 z} \vec{H}_2 = \hat{y} B e^{-jk_0 z}.$$

Using the boundary conditions gives $A = B = \frac{-J_x}{2}$.

Propagating electromagnetic waves with their electric fields pointing in the same fixed direction are called *linearly polarized* (Fig. 2.1a). Often, the end point of the electric field vector traces a circular spiral, as the wave propagates. Such an electromagnetic wave is called *circularly polarized* (Fig. 2.1b, c). A propagating electromagnetic wave can be a linear superposition of two other ones as

$$\vec{E} = \left(E_{0x}\hat{x} + E_{0y}\hat{y}\right)e^{-jk_0z}$$

A linearly polarized wave can result from $E_{0x} \neq 0$, $E_{0y} = 0$ or vice versa, or $E_{0x} \neq 0$, $E_{0y} \neq 0$, and the two electric field vectors are separated by a phase angle.

$$\phi = \arctan\left(\frac{E_{0y}}{E_{0x}}\right)$$

On the other hand, the electric field of the propagating wave can be represented as:

$$\vec{E} = (E_0\hat{x} - jE_0\hat{y})e^{-jk_0z},$$

and then the corresponding time domain electromagnetic wave is

$$\vec{E} = E_0 \cos\left(\omega t - k_0z\right)\hat{x} + E_0 \cos\left(\omega t - k_0z - \frac{\pi}{2}\right)\hat{y}.$$

The phase difference is $\frac{\pi}{2}$ and the wave equation van be re-written as

$$\vec{E} = E_0 \cos\left(\omega t - k_0z\right)\hat{x} + E_0 \sin\left(\omega t - k_0z\right)\hat{y},$$

which is the equation for a circle. The phase angle is $\phi = \omega t$—this is a *right circularly polarized* wave.

A left circularly polarized wave (Fig. 2.1c) arises when

$$\vec{E} = (E_0\hat{x} + jE_0\hat{y})e^{-jk_0z}.$$

The magnetic field associated with a circularly polarized wave is

$$\vec{H} = \frac{E_0}{\eta_0}\hat{z}x(\hat{x} - j\hat{y})e^{-jk_0z},$$

which is also right circularly polarized. Identical arguments hold for left circularly polarized electromagnetic waves.

All electromagnetic fields carry energy and consequently power (energy/time). For the steady-state case of sinusoidal electromagnetic wave, the time averaged energy stored in the electric field is $W_E = \frac{1}{4} \int \vec{E}\vec{D}dV$ or $W_E = \frac{\varepsilon}{4} \int \vec{E}\vec{E}^{PRIME} dv$.

In a similar manner, the energy stored in the magnetic field is

$$W_M = \frac{\mu}{4} \int \vec{H}\vec{H}^{PRIME} dV.$$

These expressions can be used to analyze and reason about Poynting's theorem that enables analysis of energy conservation of electromagnetic fields and sources. The total electric current density of an electric current source with conduction current density $\vec{D} = \sigma\vec{E}$ and source current density $\vec{J}_s$ is $\vec{J} = \vec{J}_S + \sigma\vec{D}$.

A little manipulation of Maxwell's equations gives

$$\vec{H}^{PRIME} \cdot \left(\nabla \times \vec{E}\right) = -j\omega\mu\left\{\vec{H}\right\}^2 - \vec{H}^{PRIME}\vec{M}_S \text{ and}$$

$$\vec{E} \cdot \left(\nabla \times \vec{H}\right) = \vec{E} \cdot \vec{J}^{PRIME} - j\omega\epsilon^{PRMR}\left\{\vec{E}\right\}^2 = \vec{E} \cdot J_S^{PRIME} + \left(\sigma - j\omega\epsilon^{PRIME}\right)\left\{\vec{E}\right\}^2$$

Combining these two expressions gives

$$\nabla \cdot \left(\vec{E} \times \vec{H}^{PRIME}\right) = \vec{H}^{PRIME} \cdot \left(\nabla \times \vec{E}\right) - \vec{E} \cdot \left(\nabla \times \vec{H}^{PRIME}\right)$$

Now applying the divergence theorem, and integrating over a volume V, gives

$$\int \nabla \cdot \left(\vec{E} \times \vec{H}^{PRIME}\right) dv = \int \vec{E} \times \vec{H}^{PRIME} \, ds$$

where S is a closed surface enclosing the volume V. This is the Poynting's theorem. This means that the power delivered by the electric source inside a volume V, enclosed by a closed surface S, is

$$P_S = \frac{-1}{2} \int \left(\vec{E} \cdot \vec{J}_S^{PRIME} + \vec{H}^{PRIME} \cdot \vec{M}_S\right) dv.$$

The Poynting vector is defined as $\vec{S} = \vec{E} \times \vec{H}$

Closely associated with the concept of energy carried by an electromagnetic wave is reflection from a good conductor, or how does a mirror work (light is, after all, electromagnetic radiation). A small fraction of the incident energy is absorbed, and the rest is reflected. With reference to Fig. 2.2, the lossless medium ends at z = 0, and

Fig. 2.2 Electromagnetic wave energy absorption at a good (not perfect) conductor

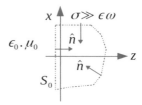

the good conductor extends from z > 0. The real average power entering the conductor is characterized by the cross-sectional surface S_S and the surface S inside the conductor:

$$P_{AVG} = \frac{1}{2} \int \left(\vec{E} \times \vec{H}^{PRIME} \right) . \hat{n} ds$$

where the integral is carried out over the combined surface $S_S + S$. Here $\hat{n}$ is the unit normal pointing into the surface. The contribution to the integral from the internal surface S can be made zero by appropriate selection of the incident wave, e.g., if the incident wave is normal to the surface, then the Poynting vector $\vec{S} = \vec{E} \times \vec{H}$ is tangential to the surface(up, down, bottom, and top). The decay of the electromagnetic wave starting from the surface of the conductor (z = 0) is very rapid, so that by making the surface end very far from z = 0, the electromagnetic wave amplitude can be made effectively zero at the surface S farthest away from z = 0. Then the integral

$$P_{AVG} = \frac{1}{2} \int \left(\vec{E} \times \vec{H}^{PRIME} \right) . \hat{n} ds$$

is really performed over the surface S_S **only**. Now,

$$\hat{z} \left(\vec{E} \times \vec{H}^{PRIME} \right) = \hat{z} \left(\vec{E} \times \vec{H}^{PRIME} \right) = \eta \vec{H} \cdot \vec{H}^{PRIME}.$$

Then the integral for the average power over the surface S_S **only** becomes $P_{AVG} = \frac{1}{\sigma \delta_s} \int \left\{ \vec{H} \right\}^2 ds$ where $\frac{1}{\sigma \delta_s}$ is the surface resistivity at the interface at z = 0.

Unlike the idealized scenarios examined so far, real world through which an electromagnetic wave propagates is lossy, characterized by complex $\in, \mu$. Let such a medium extend from z > 0. Then, for z < 0, the electric and magnetic field vectors are

$$\vec{E} = \hat{x} E_0 e^{-jk_0 z}, \vec{H} = \frac{\hat{y}}{\eta_0} E_0 e^{-jk_0 z}.$$

In the lossy medium (z > 0), the reflected electric|magnetic field vectors are

$$\vec{E} = \hat{X} \Gamma E_0 e^{jk_0 z}, \vec{H} = -\hat{y} \frac{\Gamma}{\eta_0} E_0 e^{jk_0 z}$$

where Γ is the reflection coefficient. The electromagnetic waves propagating along the positive z axis (z > 0) have electric and magnetic field vectors and magnetic field vectors defined as

$$\vec{E} = \hat{x}TE_0e^{i\gamma z}, \vec{H} = \frac{\hat{y}Te^{-\gamma z}}{\eta}$$

where T is transmission coefficient, and

$$\eta = \frac{j\omega\mu}{\gamma} \text{ and } \gamma = \alpha + j\beta = j\omega\sqrt{\epsilon\mu}\sqrt{1 - \frac{j\sigma}{\epsilon\omega}}.$$

Now, applying the boundary condition that the tangential $\gamma = j\$beta = j\omega\sqrt{\epsilon\mu} = jk_9\sqrt{\epsilon_r\mu_r}$ components of the electric, magnetic fields must be continuous at z = 0, $1 + \Gamma = T$, $\frac{1-\Gamma}{\eta_9} = \frac{T}{\eta}$, which can be manipulated to give

$$\Gamma = \frac{\eta - \eta_{0\grave{\iota}}}{\eta + \eta_0}, T = \frac{2\eta}{n + \eta_0}.$$

For a lossless medium, $\alpha = 0$ for z > 0. Then

$$\gamma = j\beta = j\omega\sqrt{\epsilon\mu} = jk_0\sqrt{\epsilon_r\mu_r}$$

and the wavelength and phase velocities are respectively

$$\lambda = \frac{2\pi}{\beta} = \frac{2\pi}{\omega\sqrt{\epsilon\mu}} = \frac{\lambda_0}{\sqrt{\epsilon_R\mu_R}} \text{ and } V_{PPHASE} = \frac{\omega}{\beta} = \frac{1}{\sqrt{\epsilon\mu}} = \frac{C}{\sqrt{\epsilon_R\mu_R}}.$$

The corresponding wave impedance is

$$\eta = \frac{j\mu\omega}{\gamma} = \sqrt{\frac{\mu}{\epsilon}} = n_0\sqrt{\frac{\mu_R}{\epsilon_R}}.$$

In this special lossless case, both $\vec{E}, \vec{H}$ in phase (coherent) and the Poynting vectors for the reflected, transmitted waves can be determined in a straightforward way. For z < 0

$$\vec{S}^{MIINUS} = \frac{\hat{z}\{E_0\}^2}{\eta_0}\left(1 - \{\Gamma\}^2 + 2j\Gamma\sin(2k_0z)\right)$$

For z > 0 the Poynting vector is

$$\vec{S}^{PLUs} = \hat{z}\frac{\{E_0\}^2\left(1 - \{\Gamma\}^2\right)}{\eta_0}$$

The time averaged power through 1 meter squared area for z < 0 is

$$P^{MIMUS} = \frac{\{E_0\}^2 \left(1 - \{\Gamma\}^2\right)}{2\eta_0},$$

and the time averaged power through 1 meter squared area in z > 0 is

$$P^{PLUS} = \frac{\{E_0\}^2 \left(1 - \{\Gamma\}^2\right)}{2\eta_0}.$$

Clearly, energy flowing across the interface in the lossless media (z < 0, z > 0) is conserved. It must be noted that if the incident reflected wave Poynting vectors for the two media are computed separately,

$$\vec{S}_{INCIDEMT} = \frac{\hat{z}\{E_0\}^2}{\eta_0} \quad \text{and} \quad \vec{S}_{REFLECTED} = \frac{-\hat{z}\{E_0\}^2\{\Gamma\}^2}{\eta_0},$$

the mismatch is evident, and indicates that the breakup of the Poynting vector into incident, reflected parts does not guarantee meaningful results.

If the region (x > 0) is occupied by a good (**not perfect**) conductor, then

$$\gamma = \alpha + j\beta = \frac{1+j}{\delta_S}$$

and the impedance is $\eta = \frac{1+j}{\sigma\delta_S}$.

For z < 0, the Poynting vector is

$$\vec{S}^{MINUS} = \frac{z\{E_0\}^2 \left(1 - \{\Gamma\}^2 + 2j\Im\left(\Gamma e^{j2k_0 z}\right)\right)}{\eta_0}.$$

Right at the interface $z = 0$, the expression for the Poynting vector reduces to

$$\vec{S}_{MINUS} = \frac{\hat{z}\{E_0\}^2}{\eta_9} \left(1 + \{\Gamma\} - \{\Gamma\}^2 - \Gamma^{PRIME}\right).$$

Similarly, for the region z > 0, the corresponding Poynting vector evaluated at $z = 0$ is the same as the Poynting vector for z < 0 evaluated at $z = 0$:

$$\vec{S}_{PLUS} = \vec{S}_{MINUS} = \frac{\hat{z}\{E_0\}^2}{\eta_9} \left(1 + \{\Gamma\} - \{\Gamma\}^2 - \Gamma^{PRIME}\right).$$

The sum of the incident and reflected Poynting vectors at the interface $z = 0$ is

$$\vec{S}_{INCIDENT} + \vec{S}_{REF} = \frac{\hat{z}\{E_0\}^2}{\eta_0}\left(1 - \{\Gamma\}^2\right).$$

The corresponding time averaged power through a 1 meter squared area for $z < 0$ and $z > 0$ are respectively

$$P_{MINUS} = \frac{\hat{z}\{E_0\}^2\left(1 - \{\Gamma\}^2\right)}{2\eta_0} \quad \text{and}$$

$$P_{PLUS} = \frac{\hat{z}\{E_0\}^2\left(1 - \{\Gamma\}^2 e^{-2\alpha z}\right)}{2\eta_0}.$$

The electric volume current density flowing through the good conductor $(z > 0)$ is

$$j = 0\{E_0\}^2\left(1 - \{\Gamma\}^2 e^{-2\alpha z}\right).$$

Consequently, the average electrical power transmitted into a 1 meter squared cross section that is inside the good conductor is

$$P_{TRANS} = \frac{\int \vec{E} \cdot \vec{J}^{PRIME} dv}{2} = \frac{\sigma\{E_0\}^2\{T\}^2}{2}\int e^{-2\alpha z} dz = \frac{\sigma\{E_0\}^2\{T\}^2}{4\alpha}.$$

For a perfect conductor, $\sigma \to \infty$ and then the electric and magnetic field vectors at $z = 0$ can be re-written as

$$\vec{E} = \vec{E}_{INC} + \vec{E}_{REF} = -\hat{x}2jE_0\sin(k_0 z) \quad \text{and}$$

$$\vec{H} = \vec{H}_{INC} + \vec{H}_{REF} = -\hat{y}\frac{2}{\eta_0}E_0\cos(k_0 z).$$

The Poynting vector, in the region $z < 0$, is:

$$\vec{S}_{MINUS} = \vec{E} \times \vec{H}^{PRIME} = \frac{\hat{z}4j\{E_0\}^2\cos(k_0 z)\sin(k_0 z)}{\eta_0}.$$

As the real part of the Poynting vector is zero, no real power is delivered by the incident electromagnetic wave to the perfect conductor. The surface current density, in the limit of infinite conductivity, evaluated at $z = 0$ is

$$\vec{J}_S = \hat{n} \times \vec{H}^{PRIME} = \frac{-\hat{Z} \times (\hat{y}2E_0 \cos(k_0z))}{\eta_9} = \frac{2E_0\hat{x}}{\eta_0}.$$

Good|perfect conductors are idealized—real-world conductors have surface impedances, conductor, and therefore need to be taken into account, to analyze and design RF|microwave circuits. To analyze this energy loss, consider a good conductor extending for $z > 0$, and a plane electromagnetic wave is incident normally on the surface. This is reflected back, but some of the wave energy is dissipated inside the conductor. From Joule's law, the transmitted power through a 1 meter squared area is

$$P_{TRANSMIT} = \frac{\int \vec{E} \cdot \vec{J}^{PRIME} dV}{2} = \frac{\int \{J\}^2 dV}{2\sigma} = \frac{\sigma\{E_0\}^2\{T\}^2}{4\alpha}.$$

This expression can be simplified again, using

$$\frac{\sigma\{T\}^2}{\alpha} = \frac{8R_S}{\eta_0^2} \quad \text{to give} \quad P_{TRANSMIT} = \frac{2\{E_0\}^2 R_S}{\eta_0^2}$$

where the surface resistance is $R_s = \sqrt{\frac{\omega\mu}{2\sigma}}$.

Alternatively, the Poynting vector can be used to compute the transmitted power:

$$P_{TRANSMIT} = \frac{2\{E_0\}^2\eta}{\{\eta + \eta_0\}^2}.$$

In case $\eta \ll \eta_0$, $P_{TRANSMIT} = \frac{2\{E_0\}^2 R_S}{\eta_0^2}$.

The volume current density through the good conductor is

$$j = \hat{x}\sigma E_0 Te^{-\gamma z},$$

and the current flow per unit width is

$$\vec{J}_S = \frac{\sigma E_0 T}{\Upsilon}$$

and in the limit

$$\frac{\sigma T}{\Upsilon} = \frac{2}{\eta_0}, \vec{J}_S = \frac{2E_0\hat{x}}{\eta_0}.$$

This is identical to the expression for surface current density in case the conductivity is infinite.

If the exponentially decaying current density obtained from Joule's law is replaced by a uniform current density extending over one skin depth, such that

$$j = \frac{\overrightarrow{J_S}}{\delta_S}, 0 < Z < \delta_S,$$

the transmitted power is

$$P_{TRANSMIT} = \frac{1}{2\sigma} \frac{\int \left\{\overrightarrow{J_s}\right\}^2}{\delta_S} \, dsdz = \frac{R_s}{2} \int \left\{\overrightarrow{J_s}\right\} ds = \frac{2R_s\{E_0\}^2}{\eta_0^2}$$

where S is a surface enclosing the conductor.

A few useful theorems as the reciprocity theorem and image theorem, which are very useful in analyzing RFlmicrowave problems, are examined in detail.

Let there be two separate sources of electric and hypothetical magnetic current $\overrightarrow{J}_1, \overrightarrow{M}_1$ and $\overrightarrow{J}_2, \overrightarrow{M}_2$ that exist that generate electric and magnetic fields $\overrightarrow{E}_1, \overrightarrow{H}_1$ and $\overrightarrow{E}_2, \overrightarrow{H}_2$, respectively.

Both sources are bound by a surface S, and Maxwell's equations are obeyed by both these sources, such that

$$\nabla \times \overrightarrow{E}_1 = -j\mu\omega\overrightarrow{H}_1 - \overrightarrow{M_1}, \nabla \times \overrightarrow{H}_1 = j \in \omega \overrightarrow{E_1} + \overrightarrow{J_1} \quad \text{and}$$
$$\nabla \times \overrightarrow{E}_2 = -j\mu\omega\overrightarrow{H}_2 - \overrightarrow{M_2},$$

$$\nabla \times \overrightarrow{H}_2 = j \in \omega \overrightarrow{E_2} + \overrightarrow{J_2}.$$

The quantity $\nabla.\left(\overrightarrow{E_1} \times \overrightarrow{H_2} - \overrightarrow{E_2} \times \overrightarrow{H_1}\right)$.

can be expanded as

$$\nabla.\left(\overrightarrow{E_1} \times \overrightarrow{H_2} - \overrightarrow{E_2} \times \overrightarrow{H_1}\right) = \overrightarrow{J}_1.\overrightarrow{E}_2 - \overrightarrow{J}_2.\overrightarrow{E}_1 + \overrightarrow{M}_1.\overrightarrow{H}_2 - \overrightarrow{M}_2.\overrightarrow{H}_1,$$

and then, after integration over a volume V gives the general form of the reciprocity theorem

$$\int \nabla.\left(\overrightarrow{E_1} \times \overrightarrow{H_2} - \overrightarrow{E_2} \times \overrightarrow{H_1}\right) dv = \int \overrightarrow{J}_1.\overrightarrow{E}_2 - \overrightarrow{J}_2.\overrightarrow{E}_1 + \overrightarrow{M}_1.\overrightarrow{H}_2 - \overrightarrow{M}_2.\overrightarrow{H}_1 dv$$

where $\overrightarrow{M}_1, \overrightarrow{M}_2$ are the fictitious magnetic current densities, added from completeness.

There are three special cases where this theorem is very useful. If the surface S does not enclose any sources, then each of

$$\vec{J}_1, \vec{H}_1, \vec{J}_2, \overrightarrow{M_2} = 0$$

and the reciprocity theorem integral reduces to

$$\int \left(\vec{E}_1 \times \vec{H}_2\right) \cdot ds = \int \left(\vec{E}_2 \times \vec{H}_1\right) \cdot ds$$

where the integral is performed over a closed contour. In case the surface S encloses a perfect conductor,

$$\hat{n} \cdot \left(\vec{E}_1 \times \vec{H}_2\right) = \hat{n} \cdot \left(\vec{E}_2 \times \vec{H}_1\right) = 0$$

using the vector identity $\hat{n} \cdot \left(\vec{E}_1 \times \vec{H}_2\right) = \left(\hat{n} \times \vec{E}_1\right)\vec{H}_2$ and $\hat{n} \times \vec{E}_1 = 0$ and then the reciprocity theorem integral becomes

$$\int \left(\vec{E}_1 . \vec{J}_2 - \vec{H}_1 . \overrightarrow{M}_2\right) dv = \int \left(\vec{E}_2 . \vec{J}_1 - \vec{H}_2 . \overrightarrow{M}_1\right) dv.$$

Physically, this means that electric field $\vec{E}_2$ caused by $\vec{J}_2$ observed at $\vec{J}_1$ is identical to electric field $\vec{E}_1$ caused by $\vec{J}_1$ observed at $\vec{J}_2$.

The third case occurs when the enclosing surface is at infinity, and then the electric and magnetic fields, and their corresponding electromagnetic waves, can be considered to be plane waves.

The image theory states that in case an electric source (and corresponding current density) is placed at a distance from a ground plane, then that ground plane can be replaced by an electric source of opposite polarity and current density of opposite polarity flowing in the opposite direction (Fig. 2.3). This interesting concept has important applications as the analysis of the versatile monopole antenna.

That is, corresponding to $\vec{J}_{SRC} = \hat{x} J_{0x}$, the image current density is $\vec{J}_{IMG} = \hat{x} J_{0x}$. If the original electric source is at a distance d from the ground plane at z = d, the image source is at z = −d, and standing waves are set up between the original source and its image. These can be expressed as

$$E_x^{STANDING} = E_0\left(e^{jk_0z} - e^{-jk_0z}\right) 0 < z < d \text{ and similarly,}$$

Fig. 2.3 An electric source near a ground plane, replaced by an image source

$$H_y^{STANDING} = \frac{-E_0}{\eta_9}\left(e^{jk_0z} + e^{-jk_0z}\right)0 < z < d¿.$$

For z > d, the electric and magnetic fields are represented as simple traveling plane waves:

$$E_x^{PLIS} = E_{00}e^{jk_0z}, H_y^{PLIS} = \frac{-E_{00}}{\eta_0}e^{jk_0z}¿.$$

While the standing waves in the region $0 < z < d$ satisfy the boundary condition $E_x = 0$, $z = 0$, the other boundary conditions that must be satisfied are $\vec{E}_x$ must be continuous at $z = d$, and $\vec{H}_y$, must be discontinuous at $z = d$. Then

$$E_{X,Z=0}^{STANDING} = E_{X,Z=0}^{PLUS} \quad \text{and} \quad \vec{J}_{STANDING,z=0} = \hat{Z} \times \hat{y}\left(H_y^{STANDING} - H_y^{PLUS}\right).$$

These two equations can be manipulated to get the amplitudes of the standing and traveling waves. After that the superposition principle can be used to get the total fields.

The uniqueness theorem plays a very important role in the analysis of electrodynamics problems. **This theorem guarantees that a given set of solutions to Maxwell's equations under a given set of boundary conditions is unique.** *It states that in a volume V filled with dissipative media and enclosed by a closed surface S, the electric and magnetic fields are uniquely determined by the source (electric, fictitious magnetic) currents inside the volume, and the tangential components of the same electric and magnetic fields on the surface S. This is proved by verifying that two solutions to Maxwell's solutions in a given volume V (filled with dissipative medium) are identical, i,e.,* $\vec{E}_1 = \vec{E}_2 \wedge \vec{H}_1 = \vec{H}_2,$. So, Poynting's theorem in this volume, with conductivity zero, gives

$$\int \left(\vec{E}_1 - \vec{E}_2\right) \times \left(\vec{H}_1 - \vec{H}_2\right)^{PRIME} ds + j\omega \int \left(\epsilon\left\{\left\{\vec{E}_1 - \vec{E}_2\right\}^2 + \mu\vec{H}_1 - \vec{H}_2\right\}^2\right)dv$$

$$= 0$$

where $\vec{S}$ is a closed contour. Under electric wall $\hat{n} \times \vec{E} = 0$ or magnetic wall $\hat{n} \times \vec{H} = 0$ or fixed tangential electriclmagnetic field conditions $\hat{n} \times \vec{E} = \vec{E}_{TANGET}\hat{n} \times \vec{H} = \vec{H}_{TANGET}$ or a combination, the first part of the above integral vanishes.

If both the electric permittivity and magnetic permeability are complex valued, then   one   part   of   the   equation   is
$\int\left(\epsilon_{IMG}\left\{\vec{E_1} - \vec{E_2}\right\}^2 + \mu_{IMG}\left\{\vec{H_1} - \vec{H_2}\right\}^2\right)dv = 0$. This condition is satisfied if and only if $\vec{E}_1 = \vec{E}_2 \wedge \vec{H}_1 = \vec{H}_2$.

2.2 Transmission Line Fundamentals

Transmission line concepts [42] are essential to circuit analysis at operating frequencies of 100 s of MHz to 10s of GHz. **This is because at these ultrahigh frequencies (100 s of MHz–10s of GHz), the physical dimensions of the circuit elements are a fraction\multiple of the electrical wavelength of the electromagnetic wave flowing through the circuit.** *A transmission line is a distributed electrical network, for which currents\voltages vary over its length.*

A typical transmission line is shown in Fig. 2.4. Here, C, G, L, and R are respectively the shunt capacitance per unit length, shunt conductance per unit length, series inductance per unit length, and series resistance per unit length. The shunt capacitance is die to the two dielectric separated conductors of the transmission line, the series inductance is the total inductance die to the conductors, the series resistance represents the resistance of the finite length conductors, and the shunt conductance is due to the dielectric medium separating the conductors. Applying Kirchhoff's laws,

$$v(z,t) - R\Delta z i(z,t) - L\Delta z \frac{\partial i(z,t)}{\partial t} - v(z + \Delta z, t) = 0$$

and

$$i(z,t) - G\Delta z v(z,t) - C\Delta z \frac{\partial v(z + \Delta z, t)}{\partial t} - i(z + \Delta z, t) = 0.$$

These two equations, after some manipulation, give the steady-state equations

$$\frac{dV(z)}{dz} = -(R + j\omega L)I(z) \quad \text{and}$$

$$\frac{dV(z)}{dz} = -(C + j\omega C)V(z).$$

Defining $\gamma = \alpha + j\beta = \sqrt{(R + j\omega L)(G + j\omega C)}$ and applying separation of variables, the decoupled equations are

Fig. 2.4 Transmission line schematic and equivalent electrical circuited

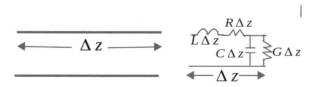

$$\frac{d^2V(z)}{dz^2} - \gamma^2 V(z) = 0 \quad \text{and} \quad \frac{d^2I(z)}{dz^2} - \gamma^2 I(z) = 0.$$

The general solutions to these equations are

$$V(z) = V_0^{PLUS}e^{-\gamma z} + V_0^{MINUS}e^{\gamma z} \quad \text{and}$$

$$I(z) = I_0^{PLUS}e^{-\gamma z} + I_0^{MINUS}e^{\gamma z}$$

where the PLUS/MINUS superscripts indicate electromagnetic waves traveling in the positivelnegative directions, respectively, A little manipulation of the above equations gives the relation between current and voltage in the transmission line:

$$I(z) = \frac{\gamma}{R + j\omega L}\left(V_0^{PLUS}e^{-\gamma z} - V_0^{MINUS}e^{\gamma z}\right)$$

which immediately leads to the definition of the **impedance per unit length, the characteristic impedance**

$$Z_9 = \sqrt{\frac{R + j\omega L}{G + j\omega C}}.$$

With this definition, the currents and voltages in the transmission line can be written as $Z_0 = \frac{V_0^{PLUS}}{I_0^{PLUS}} = \frac{-V_0^{MINUS}}{I_0^{MINUS}}$.

Combining these results, the time domain expression for the voltage wave is

$$V(z,t) = \left\{V_0^{PLUS}\right\}\cos\left(\omega t - \beta z + \phi^{PLUS}\right)e^{-\alpha z} + V_0^{MINUS}\right\}\cos\left(\omega t - \beta z - \phi^{MINUS}\right)$$

where ϕ^{PLUS}, ϕ^{MINUS} are the phase angles. In each of the expressions above, α, β are the attenuation and propagation constants, respectively. The wavelength and phase velocities of the traveling electromagnetic wave are $\lambda = \frac{2\pi}{\beta}$, $V_{PHASE} = \frac{\omega}{\beta} = f\lambda$.

The transmission line equations and expressions can be derived from Maxwell's eqs. A 1-m-long section of transmission line with cross-sectional area S and currentsl voltages between the conductors $I_0e^{\pm j\beta z}$, $V_0e^{\pm j\beta z}$, has time averaged magnetic energy

$$W_{MAG} = \frac{\mu}{4}\int \vec{H} \cdot \vec{H}^{PRIME}\, ds$$

stored in it. The evaluated integral gives $W_{MAG} = \frac{L\{I_0\}^2}{4}$, so that the inductance per unit length is $L = \frac{\mu}{\{I_0\}^2}\int H \cdot \vec{H}^{PRIME}\, ds\frac{H}{m}$.

Using identical reasoning for energy stored in the electric field,

$$W_{ELEC} = \frac{\epsilon}{4} \int \vec{E} \cdot \vec{H}^{PRIME} \, ds = C \frac{\{V_0\}^2}{4}.$$

Therefore, the capacitance per unit length is

$$C = \frac{\epsilon}{\{V_0\}^2} \int \vec{E} \cdot \vec{H}^{PRIME} \, ds \, \frac{F}{m}.$$

The energy|power loss per unit length of the conductor, die to the finite conductivity of the conductor, is

$$P_{LOSS,CONDUCTOR} = \frac{R_S}{2} \int \vec{H} \cdot \vec{H}^{PRIME} \, dl$$

where the integration is computed over both the conductors and R_S is the sheet resistance of the conductor material. From circuit theory,

$$P_{LOSS} = \frac{R\{I\}^2}{2}$$

and applying this, the resistance per unit length is

$$R = \frac{R_S}{\{I_0\}^2} \int \vec{H} \cdot \vec{H}^{PRIME} \, dl \, \frac{\Omega}{m}.$$

Using identical reasoning for the lossy dielectric in between the conductors,

$$G = \frac{\omega \epsilon_{IMG}}{\{V_0\}^2} \int \vec{E} \cdot \vec{E}^{PRIME} \, ds \, \frac{S}{m}.$$

Using these expressions, the four parameters for common transmission line as coaxial cable can be evaluated in a straightforward manner.

A lossless terminated transmission line has a load impedance connected between the two conductors at the output port, and the signal is fed in at the input port. The total voltage on the transmission line is the sum of the incident and reflected waves:

$$V(z) = V_0^{PLUS} e^{-j\beta z} + V_0^{MINUS} e^{j\beta z}$$

And the current through the transmission line is

$$I(z) = \frac{V_0^{PLUS} e^{-j\beta z}}{Z_0} - \frac{V_0^{MINUS} e^{j\beta z}}{Z_0}.$$

The current and voltage at the load are related through the load impedance:

$$Z_L = \frac{V(0)}{I(0)} = \frac{V_0^{PLUS} + V_0^{MINUS}}{V_0^{PLUS} - V_0^{MIMUS}} Z_0,$$

so that the amplitude of the reflected voltage wave is:

$$V_0^{MINUS} = \frac{Z_L - Z_0}{Z_L + Z_0} V_0^{PLUS},$$

which leads immediately to the definition of the reflection coefficient

$$\Gamma = \frac{V_0^{MINUS}}{V_0^{PLUS}} = \frac{Z_L - Z_0}{Z_L + Z_0}.$$

Now the voltage and current waves on the transmission line can be re-written as

$$V(z) = V_0^{PLUS}\left(e^{-j\beta z} + \Gamma e^{j\beta z}\right) \quad \text{and}$$

$$I(z) = \frac{V_0^{PLUS} e^{-j\beta z}}{Z_0} - \Gamma \frac{V_0^{PLUS} e^{j\beta z}}{Z_0}.$$

Both the current and the voltage waves on the transmission line consist of a linear superposition of incident and reflected waves—standing waves. The time averaged power flow along the transmission line is

$$P_{TIME,AVG} = \frac{\{V_0^{PLUS}\}^2}{Z_8} \Re\left(1 + \Gamma e^{2j\beta z} - \Gamma^{PRIME} e^{-2j\beta z} - \{\Gamma\}^2\right).$$

As the two middle terms in the expression are purely imaginary, they cancel, and the expression for the time averaged power simplifies to

$$P_{TME,AVG} = \frac{\{V_0^{PLLS}\}^2}{Z_8}(1 - \{\Gamma\}).$$

When the reflection coefficient is 0, all incident power is delivered to the load, and vice versa. Using this expression, the return loss of a transmission line is

$$RL = -20 \ln\left(\{\Gamma\}\right) dB.$$

The voltage on the transmission line is not constant:

$$\{V(z)\} = \{V_0^{PLUS}\}\{1 + \Gamma e^{-2j\beta 1}\},$$

and consequently, the maximum and minimum values for the voltage occur when the phase term's magnitude is $+/-$ 1. Consequently, the voltage standing wave ratio is defined as

$$VSWR = \frac{1 + \{\Gamma\}}{1 - \{\Gamma\}}.$$

The distance between two consecutive maximalminima is $\frac{\lambda}{2}$, and the distance between a maxima and a minima is $\frac{\lambda}{4}$.. The reflection coefficient decreases exponentially with distance along the transmission line as

$$\Gamma(1) = \frac{V_0^{MINUS} e^{-j\beta l}}{V_0^{PLUS} e^{j\beta l}} = \Gamma(0) e^{-2j\beta l}$$

This key expression allows the estimation of the reflection coefficient as a function of the length of the transmission line. Therefore, the impedance, looking into a transmission line, is

$$Z_{INPUT} = \frac{e^{j\beta i} + \Gamma e^{-j\beta l}}{e^{j\beta l} - \Gamma e^{-j\beta l}} Z_0 = \frac{1 + \Gamma e^{-2j\beta l}}{1 - \Gamma e^{-2j\beta l}} Z_0.$$

This expressions can be re-written in a more convenient, usable form (in terms of the characteristic, load impedances) as

$$Z_{INPUT} = \frac{Z_L + jZ_0 \tan(\beta l)}{Z_0 - jZ_L \tan(\beta l)} Z_0.$$

Lossless transmission lines are special cases of general lossy transmission lines. If the transmission line is short circuited at the load end,

$$Z_{INPUT} = jZ_0 \tan(\beta l)$$

which is purely imaginary. For the infinite load impedance case, the reflection coefficient is unity, and then current and voltage waves traveling along the transmission line are

$$I(z) = \frac{-2jV_0^{PLUS}}{Z_0} \sin(\beta z) \quad \text{and}$$

$$V(z) = 2V_0^{PLUS} \cos(\beta z)$$

so that the input impedance is $Z_{INPUT} = -jZ_0 \cot(\beta l)$.

If the length of the transmission line is half-wavelength, $Z_{INPUT} = Z_L$ and if it is quarter wavelength long, $Z_{INPUT} = \frac{Z_0^2}{Z_L}$.

If the lossless transmission line is infinitely long, or is terminated with the characteristic impedance, such that input impedance is Z_{INPUT}, the reflection coefficient is $\Gamma = \frac{Z_{INPUT} - Z_0}{Z_{INPUT} + Z_0}$.

In this case, a fraction of the incident wave is transmitted, and the remainder is reflected. The transmission coefficient can be expressed in terms of the reflection coefficient as

$$T = 1 + \{\Gamma\} = \frac{2Z_0}{Z_{INPUT} + Z_0}.$$

and the corresponding insertion loss is $IL = -20 \ln (\{T\}) dB$.

The lossless transmission line examined so far is an idealization of the real-world lossy transmission line. The general expression for the complex propagation constant is $\gamma = \sqrt{(R + j\omega L)(G + j\omega C)}$.

This can be simplified as $\gamma = j\omega \sqrt{CL} \sqrt{1 - j\left(\frac{R}{\omega L} + \frac{G}{\omega C}\right) - \frac{GR}{\omega^2 CL}}$.

In case both conductor and dielectric losses are small,

$$R \ll \omega L, G \ll \omega C$$

the third term in the above square root vanishes, giving

$$\gamma = j\omega \sqrt{CL} \sqrt{1 - j\left(\frac{R}{\omega L} + \frac{G}{\omega C}\right)}.$$

The second square root can be expanded in a Taylor's series expansion giving

$$\gamma = j\omega \sqrt{CL} \left(1 - \frac{j}{2}\left(\frac{R}{\omega L} + \frac{G}{\omega C}\right)\right).$$

Consequently, the approximate expressions for the attenuation and propagation constants are

$$\alpha = \frac{1}{2}\left(\frac{R}{Z_0} + GZ_0\right), \beta = \omega \sqrt{CL},$$

where the approximate value of the characteristic impedance is

$$Z_0 = \sqrt{\frac{L}{C}}.$$

The propagation constant is frequency dependent for a lossy transmission line, and the dependency is nonlinear. So the phase velocity is different for different values of the signal frequency. **This means that the different components of a wide band signal will travel down the transmission line with different frequencies, and arrive at the receiver separated out at different times.** *This is signal distortion, or pulse broadening.* Pulse broadening can be a very serious problem for long transmission lines carrying wide band signals. However, by design and construction, pulse broadening can be minimized if the following condition is satisfied by the unit length capacitance, conductance, inductance, and resistance: $\frac{R}{L} = \frac{G}{C}$.

The complex attenuation constant then becomes $\gamma = R\sqrt{\frac{C}{L}} + j\omega\sqrt{CL}$.

For a terminated lossy transmission line with load reflection coefficient Γ, the current and voltage waves are

$$I(z) = \frac{V_0^{PLUS}}{Z_0}(e^{-\gamma z} - \Gamma e^{\gamma z}) \quad \text{and}$$

$$V(z) = V_0^{PLUS}(e^{-\gamma z} + \Gamma e^{\gamma z}).$$

The reflection coefficient at a distance l from the load is $\Gamma(l) = \Gamma(0)e^{-2\gamma l}$. Therefore, the input impedance at a distance l from the load impedance is

$$Z_{INPUT} = \left(\frac{Z_L + Z_0 \tanh(\gamma l)}{Z_0 + Z_L \tanh(\gamma l)}\right) Z_0.$$

The input power delivered at the transmission line input is

$$P_{INPUT} = \frac{\{V_0^{PLUS}\}^2}{Z_0}\left(1 - \{\Gamma\}^2\right)e^{2\alpha l}$$

and the power delivered to the load is:

$$P_{LOAD} = \frac{\{V_0^{PLUS}\}^2}{Z_0}\left(1 - \{\Gamma\}^2\right).$$

In a straightforward manner, the power loss is

$$P_{LOSS} = P_{INPUY} - P_{LOAD} = \frac{\{V_0^{PUS}\}^2}{Z_0}\left(e^{2\alpha|} - 1 + \{\Gamma\}^2\left(1 - e^{-2\alpha|}\right)\right).$$

Two very important schemes for calculating transmission line signal attenuation are the perturbation method and Wheeler's incremental inductance rule.

Fig. 2.5 Single microstrip transmission line and its electric and magnetic fields

Of the various types of transmission line commonly used in RF|microwave circuits, the microstrip is the most popular, because of ease of design and fabrication. In its simplest form, it consists of a metal line of width W, etched on a very thin grounded dielectric layer of relative dielectric constant ϵ_r and thickness d. Figure 2.5 shows the electric field lines for the microstrip.

As the dielectric layer does not cover the microstrip (unlike a strip line), the discontinuity complicates the analysis. **For the microstrip, most of the electric field lines are embedded in the dielectric later, and a small fraction is in the air so that the phase velocities in the dielectric and the air do not match. To compensate for this mismatch, the waves in a microstrip are termed quasi-TEM.** Then good approximations to the phase velocity and characteristic impedance can be obtained from quasi-static analysis. So the phase velocity and propagation constant can be expressed as

$$V_P = \frac{C}{\sqrt{\epsilon_{eff}}}, \beta = k_0\epsilon_{eff}, 1 < \epsilon_{eff} < \epsilon_r,$$

where c is the speed of light in vacuum. The common expressions used in microstrip design are listed as follows:

$$\epsilon_{eff} = \frac{\epsilon_r + 1}{2} + \frac{\epsilon_r - 1}{2\sqrt{1 + \frac{12d}{W}}}.$$

and the expressions for the characteristic impedance are

$$Z_0 = \frac{60}{\sqrt{\epsilon_{eff}}} \ln\left(\frac{8d}{W} + \frac{W}{4d}\right), \frac{W}{d} \leq 1 \text{ and}$$

$$Z_0 = \frac{120\pi}{\sqrt{\epsilon_{eff}}\left(1.393 + \frac{W}{d} + 0.667\left(1.444 + \frac{W}{d}\right)\right)}, \frac{W}{d} > 1$$

For a given value for the characteristic impedance, the ratio of the microstrip to the dielectric layer thickness can be expressed analytically. The difficulty is that these expressions are very complicated, and which one to use depends on the ratio

$$\frac{W}{d} > 2 \ \text{or} \frac{W}{d} < 2.$$

Consequently, the use of these expressions is a trial-error method, and in a real-world microstrip fabrication case, the width and dielectric thickness are predefined. The attenuation die to dielectric loss is

$$\alpha_{DIELECTRIC} = \frac{k_0 \epsilon_r (\epsilon_{\text{eff}} - 1) \tan(\delta)}{2\sqrt{\epsilon_{\text{eff}}}(\epsilon_r - 1)},$$

where $\tan(\delta)$ is the loss tangent. Similarly, the attenuation die to conductor loss is

$$\alpha_{CONDUCTOR} = \frac{\frac{\mu_0 \omega}{2\sigma}}{Z_0 W} \frac{Np}{m}.$$

Electromagnetic waves propagate in three modes, TEM (transverse electromagnetic), TE (transverse electric), and TM (transverse magnetic). The TEM mode is characterized by $E_z = H_z = 0$ where z is the propagation axis|direction. The TE mode is identified by $E_z = 0$, $H_z \neq 0$ and the TM mode is characterized by $E_z \neq 0$, $H_z = 0$.

Coupled microstrip lines are very common in RF|microwave circuits, and these interact with each other depending on whether the signal is flowing through them in the same (even) or opposite (odd) directions. Assuming a quasi-TEM propagation mode, the characteristics of the coupled lines can be determined from the mutual capacitances and phase velocities. The capacitances that arise from the coupled lines are

C_{11}, C_{22} capacitance of each microstrip with the ground plane below.

C_{12} capacitance of the two microstrips in the absence of the ground plane.

For even mode of propagation, the electric field is symmetric about the centerline in between the microstrips, while in the odd propagation mode, the electric field lines have odd symmetry about the centerline in between the microstrips, and an electric field exists between them (Fig. 2.6a, b). The equivalent capacitances and characteristic impedances of the two configurations are

Fig. 2.6 (**a**) Coupled microstrip lines' even mode. (**b**) Coupled microstrip lines' odd mode

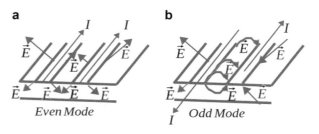

$$C_{EQ,EVEN} = C_{11} = C_{22}Z_{0,EVEN} = \sqrt{\frac{L}{C_{EQ,EVEN}}} = \frac{1}{V_{PHASE}C_{EQ,EVEN}}.$$

For the odd propagation case, the corresponding expressions for the equivalent capacitance and characteristic impedance are

$$C_{EQ,ODD} = C_{11} + 2C_{12} = C_{22} + 2C_{12}C_{EQ,ODD} = \frac{1}{V_{PHASE}C_{EQ,ODD}}.$$

The even|odd characteristic impedances $Z_{0,\ EVEN} \vee Z_{0,\ ODD}$ represent the characteristic impedance of one of the coupled microstrips when the signal is flowing through them in the same|opposite directions. Any arbitrary excitation|signal pattern of a set of parallel microstrips can be examined as a linear combination of the coupled even|odd propagation mode modes. Quasi-static techniques are used to compute the values for the coupling capacitors, etc.

Commonly available nomograms are used to determine the coupled characteristic impedances of the even|odd coupled microstrips.

The even|odd mode coupled microstrip analysis is used to determine the

properties of the directional coupler (Fig. 2.7a, b, c and d), the simplest planar transformer, used to couple RF|microwave signals from one microstrip to another, at RF|microwave frequencies. This is a four-port network, whose three ports are terminated with the characteristic impedance. Port 1 is the input port, 2 the through port, 3 the output, and port 4 the isolated port. Both conductors share a common ground plane.

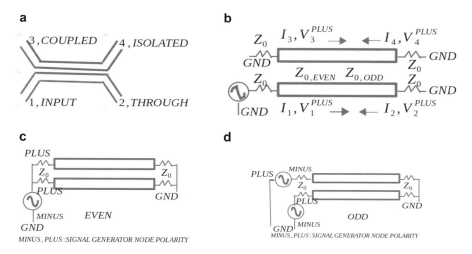

Fig. 2.7 (**a**) Directional coupler. (**b**) Equivalent electrical circuit for directional coupler. (**c**) Coupled microstrip even propagation mode. (**d**) Coupled microstrip transmission line odd propagation mode

Using even mode propagation analysis methods,

$$I_1^E = I_3^E, I_4^E = I_2^E, V_1^E = V_2^3, V_4^E = V_2^E$$

and similarly for the odd mode propagation mode analysis results,

$$I_1^o = -I_3^o, I_4^o = -I_2^o, V_1^o = -V_2^o, V_4^o = -V_2^o.$$

The input impedance, therefore, is $Z_{INPUT} = \frac{V_1}{I_1} = \frac{V_E + V_O}{I_E + I_O}$.

In a straightforward manner, the even|odd input impedances are

$$Z_{INPUT}^E = Z_{0,E} \frac{Z_0 + jZ_{0,E} \tan(\theta)}{Z_{0,E} + jZ_0 \tan(\theta)}$$

and similarly,

$$Z_{INPUT}^0 = Z_{0,o} \frac{Z_0 + jZ_{0,o} \tan(\theta)}{Z_{0,o} + jZ_0 \tan(\theta)}.$$

For each mode, the line looks like transmission line of characteristic impedance $Z_{0, E} \vee Z_{0, o}$ terminated in an impedance Z_0. Using simple voltage divider concept, the even|odd currents|voltages at input node 1 are

$$I_1^E = \frac{V_0}{Z_{INPUT}^E + Z_0} \quad \text{and} \quad V_1^E = \frac{V_0 Z_{INPUT}^E}{Z_{INPUT}^E + Z_0}$$

$$I_1^O = \frac{V_0}{Z_{INPUT}^O + Z_0} \quad \text{and} \quad V_1^O = \frac{V_0 Z_{INPUT}^0}{Z_{INPUT}^O + Z_0}.$$

Using these expressions, the input impedance can be re-written as

$$Z_{INPUT} = Z_0 + \frac{2(Z_{INPUT}^E Z_{INPUT}^o - Z_0^2)}{Z_{INPUT}^E + Z_{INPUT}^o + Z_0}.$$

In case $Z_0 = \sqrt{Z_{INPUT}^E Z_{INPUT}^O}$, $Z_{INPUT} = Z_0$., the voltage at node 3 is

$$V_3 = V\left(\frac{Z_{INPUT}^E}{Z_{INPUT}^E + Z_o} - \frac{Z_{INPUT}^o}{Z_{INPUT}^o + Z_0}\right).$$

Using

$$\frac{Z_{INPUT}^E}{Z_{INPUT}^E + Z_0} = \frac{Z_0 + jZ_0^R \tan(\theta)}{2Z_0 + j(Z_0^E + Z_0^o) \tan(\theta)}$$

and a very similar expression for the odd input impedance, the expression for the voltage at node 3 reduces to

$$V_3 = \frac{jV \tan (\theta) \left(Z_0^E - Z_0^o\right)}{2Z_0 + j\left(Z_0^E + Z_0^o\right) \tan (\theta)}.$$

The coupling coefficient is defined as

$$C = \frac{Z_0^E - Z_0^o}{Z_0^E + Z_0^o}, \text{ and recognizing that}$$

$$\sqrt{1 - C^2} = \frac{2Z_0}{Z_0^E + Z_0^o},$$

the expression for the voltage at node 3 becomes:

$$V_3 = \frac{jCV \tan (\theta)}{\sqrt{1 - C^2} + j \tan (\theta)}.$$

In a similar fashion,
$V_2 = \frac{V\sqrt{1-C^2}}{\sqrt{1-C^2} \cos (\theta) + j \sin (\theta)}$ and $V_4 = 0$. In each of these expressions, θ is the electrical length expressed in radians. At very low frequencies $\left(\theta \ll \frac{\pi}{2}\right)$,, very little power is coupled and the output at port 3 is negligible. All signal power appears at port 2. The optimum operating point for a small size coupler with minimum loss is $\theta = \frac{\pi}{2}$ when all input signal power input at port 1 is available at port 3. These maxima occur at odd multiples of $\frac{\pi}{2}$.

When $\theta = \frac{\pi}{2}$ and the length of the coupler is $\frac{\pi}{4}$,

$$\frac{V_3}{V} = C, \frac{V_2}{V} = -j\sqrt{1 - C^2}.$$

Clearly, $C < 1$ is the voltage coupling factor at the design frequency $\theta = \frac{\pi}{2}$. All input power is available at the output

$$P_{INPUT} = \frac{\{V\}^2}{2Z_0}, P_2 = \frac{\sqrt{1 - C^2}\{V\}^2}{2Z_0}, P_3 = \frac{\{C\}^2\{V\}^2}{2Z_0}$$

and $P_4 = 0$, so that $P_{INPUT} = P_2 + P_3 + P_4$.

Given values for the characteristic impedance and the coupling factor, the even/odd characteristic impedances can be expressed as $Z_{0,E} = \sqrt{\frac{1+C}{1-C}}$ and $Z_{0,o} = \sqrt{\frac{1-C}{1+C}}$.

The parallel transmission line-based directional coupler is the key to planar spiral inductor transformer, and can be used also to curtail cross talk between traces on ultrahigh-frequency signal printed circuit boards.

2.3 Inductance Concepts

Inductance [1–42] is the property of an electrical conductor arising from its ability to store energy in its magnetic fields when current flows through it. This property is exploited in a CL tank oscillator, where theoretically (ideal CL tank does not have any energy losses) there is no energy loss, and energy moves back and forth in between the electric field of the capacitor and the magnetic field of the inductor. From transmission line theory [42] (examined earlier), the inductance of the transmission line inductor is

$$L = \frac{\mu}{I_0} \int \vec{H} \cdot \vec{H}^{PRIME} ds \frac{H}{m}.$$

From circuit theory, the impedance of an electrical conductor is $Z_{COND} = R_{COND} + j\omega L_{COND}$ where L_{COND}, R_{COND} are respectively the inductance and Ohmic resistance of the conductor. Both the inductance and Ohmic resistance of the of the conductor are independent of the signal frequency; they depend on the physical dimensions and physical properties of the inductor (magnetic permeability, resistivity).

Inductance arises because an electrical conductor opposes a change in the electric current flowing through it. Current flowing through an electric conductor creates a magnetic field around it. The magnetic field strength depends on the magnitude of the current, and tracks changes in current that caused the magnetic field. Faraday's law of induction states that any change in magnetic field through a circuit induces an electromotive force (EMF—voltage) in the conductor, through electromagnetic induction. This induced voltage created by the changing current opposes the change causing it, as per Lenz's law, and the induced electromotive force is called the **back EMF**.

Inductance is defined as the ratio of the induced voltage to the rate of change of current causing it, basically a proportionality factor that depends on the geometry of circuit conductors and physical properties of surrounding materials as magnetic permeability

According to Ampere's current law and Maxwell's equations, current flowing through a conductor generates a magnetic field around it. The total magnetic flux through a circuit is the product of the orthogonal component of the magnetic flux density and the area of the surface that completely encloses the current path. For a periodic time-variant current, the magnetic flux (and correspondingly the magnetic flux) generated by this current will also periodically vary with time. **So, Faraday's**

law of induction states that any change in flux through a circuit induces an electromotive force (EMF—voltage) in the circuit, proportional to the time rate of change of flux.

$E \frac{-a\phi}{dt}$ *where the negative sign is significant as it indicates that the induced electromotive force is in a direction opposite to the inducing field*, hence the term **back EMF**. If the current is increasing, the voltage is positive at the conductor end through which the current enters and negative at the end through which it leaves, reducing the current. Conversely, if the current is decreasing, the voltage is positive at the end through which the current leaves the conductor, tending to maintain the current. Self-inductance, usually called inductance, is the ratio between the induced voltage and the rate of change of the current:

$$V(t) = L\frac{di}{dt}.$$

Clearly, inductance is a property of a conductor, due to its magnetic field, which opposes changes in current through the conductor. The SI unit of inductance is the henry (H).

All conductors have some inductance. This inductance may enhance or degrade the performance of practical electrical devices. The inductance of a circuit depends on the geometry of the current path, and on the magnetic permeability of nearby materials. For example, ferromagnetic materials with a higher permeability near a conductor tend to increase the magnetic field and inductance. Any alteration to a circuit which increases the magnetic flux through the circuit produced by an inducing current increases the inductance, as **inductance is also equal to the ratio of magnetic flux to current:**

$$L = \frac{\phi(i)}{i}.$$

An inductor consists of a conductor shaped to increase the magnetic flux, to add inductance to a circuit. The most common inductor is a wire shaped as a coil or helix. A coiled wire has a higher inductance than a straight wire of the same length, since the magnetic field lines pass through the circuit multiple times, resulting in multiple flux linkages.

The inductance of a coil can be increased by inserting a piece of ferromagnetic material in the hole in the center. The magnetic field of the coil magnetizes the material of the core, aligning its magnetic domains, and the magnetic field of the core adds to that of the coil, increasing the flux through the coil.

When multiple inductors are located close to each other, the magnetic field of one can pass through the other, resulting in **inductive coupling**. From Faraday's law of induction, a change in current in one inductor induces a change in magnetic flux in another inductor and a voltage. This fact immediately extends the concept of inductance by defining the **mutual inductance** of two inductors as the ratio of voltage induced in an inductor to the rate of change of current in the inductor. This is the working principle of a transformer.

An increasing current through an inductor with inductance L induces a voltage across that inductor with a polarity opposite to the inducing current. Also a voltage drop, due to the inductor's resistance, is also present, but will be very small due to the very low resistivity of most conductors. Therefore, charges flowing through the inductor lose potential energy, and the energy supplied by the external power|signal supplies the energy to overcome this potential barrier. This supplied energy from the external stored in the inductor's magnetic field that encloses the inductor. At any given time, the power flowing into the magnetic field, which is equal to the rate of change of the stored energy $E(t)$, is the product of the current and voltage across the inductor:

$$P(t) = i(t)v(t) = \frac{dE}{dt} = iL\frac{di}{dt}.$$

When there is no current, there is no magnetic field and the stored energy is zero. Ignoring very small resistive losses, **the energy (unit Joules) stored by an inductor with a current through it is equal to the amount of work required to establish the current through the inductance from zero, and therefore the magnetic field**:

$$E = \int iLdi = L\int idi = \frac{Li^2}{2}.$$

The concept of mutual inductance has been introduced earlier, and the very interesting case of two loops is examined. The two loops are independent closed circuits that can have different lengths, and any orientation in space, and carry different currents. The error terms, which are often not fully taken into account, are only small if the geometries of the loops are smooth and convex, and they do not have too many kinks|sharp corners, crossovers, parallel segments, concave cavities, or other geometrical deformations. Also, the loops must be made of thin wires such that the radius of the wire is negligible compared to its length:

$$L = \frac{\mu_0}{4\pi} \frac{\int dx_1 \cdot dx_2}{\{x_1 - x_2\}} ¿.$$

The inductors that will be examined in detail are all planar [1–39], and are called "printed inductors" as they are fabricated using the same technology used for making printed circuit boards and integrated circuits. The commonest planar printed inductor is the rectangular strip inductor—all planar spiral inductors are made by sequentially connecting planar rectangular strip inductors. At RF|microwave frequencies, these rectangular strip inductors are nothing more than microstrip transmission lines fabricated inside an integrated circuit or a printed circuit board. The analysis for a planar spiral inductor inside an integrated circuit is more complicated than a similar inductor fabricated on a printed circuit board, because of unavoidable parasitic capacitances arising from the coupling of the inductor conductor with the doped substrate and silicon oxide insulating layer below the metal layer. The other common planar microstrip loop inductor is the circular loop.

The first detailedIsystematic analysis of planar rectangular inductor was done by Grover [15], later extended to planar spiral inductors by Greenhouse [14]. Grover's formula for a rectangular cross section (height b, width a. and length l) is

$$L = \frac{\mu_0 l}{2\pi} \left(0.5 + \ln\left(\frac{2l}{a+b}\right) - 0.2235 \ln\left(\frac{a+b}{l}\right) \right).$$

Clearly, the inductance is directly proportional to the length. A number of others have proposed variants of this expression, but these suffer from a number of issues.

Almost all inductance expressions are derived from the key expression for the impedance of an inductor:

$$L = \frac{\vec{\jmath}Z}{\omega} = \frac{\mu_0}{\pi l^2} \frac{\int \vec{J}\left(\vec{y}\right) J(x)^{PRIME} dv_1 dv_2}{\rho_{xy}}$$

where x, y are the current source and observation locations and

$$\rho = \sqrt{(x_1 - x_2)^2 + (y_1 - y_2)^2 + (z_1 - z_2)^2}.$$

The inductance of a rectangular microstrip of width w, dielectric thickness h, and length l is

$$L = \frac{601}{c} \left(\frac{8h}{w} + \frac{w}{4h}\right) H, \frac{w}{h} < 1 \text{ and}$$

$$L = \frac{120\pi l}{c} \left(\frac{1}{1.393 + 0.667 \ln\left(1.444 + \frac{w}{h}\right) + \frac{w}{h}}\right) H \text{ for } \frac{w}{h} \geq 1$$

The inductance calculated is the external inductance of the region bounded by the flat rectangular inductor and the ground plane.

In a similar fashion, the inductance of a square loop inductor is

$$L_{SQUARE} = \frac{2N^2 \mu_0 \mu_r w}{\pi} \left(\ln\left(\frac{w}{a}\right) - 0.7724\right)$$

where N is the number of turns, w the width of each side, a the conductor thickness, and μ_r the relative magnetic permeability of the substrate material. The inductance of a circular loop is

$$L_o = \mu_0 a \left(\ln \left(\frac{8a}{r} \right) - 2 \right)$$

where a is the loop radius and r the conductor width.

The concepts and formalism introduced by Grover [15] were extended by Greenhouse [14] who **analyzed the planar spiral inductor as a series connected set of rectangular planar inductors, and included mutual inductances between adjacent, parallel inductors with current flowing through them in both same and opposite directions.**

Two key concepts must be understood before examining Greenhouse's formulation for the planar spiral inductor. Of these two, geometric mean distance (GMD) is very important.

The geometric mean of a set of N numbers $(n_1, \ldots .n_N)$ is
$GM = \sqrt[N]{n_1 \cdot n_2 \ldots n_N}$. This concept is used to calculate the GMD and GMR. In GMD the geometrical mean of distances between the conductors of a transmission line is calculated. In the most general case, if there are two arbitrary transmission lines with m and n conductors, respectively, then the geometric mean distance between them is

$$GMD = \sqrt[mn]{(d_{11}d_{13}..d_{1m}) \ldots (d_{n1}d_2..d_m)}.$$

Greenhouse's scheme to compute the total inductance of a planar spiral inductor takes into account the self-inductance of each segment, as well as the mutual inductance of each parallel (nearest neighbor and otherwise) pair of inductor segments, where each segment is rectangular. Therefore, the total inductance of a planar spiral inductor is

$$L_{TOTAL} = \sum L_{i,SELF} + \sum L_{MUTUAL}.$$

To compute the mutual inductance of any two parallel rectangular inductor segments, the following two rules apply:

- If the current flowing in each of the pair is in the same direction, the mutual inductance of this pair is added to the sum of self-inductances.
- If otherwise, the current through each segment of the pair is in opposite directions, the mutual inductance of this pair is subtracted from the sum of self-inductances.

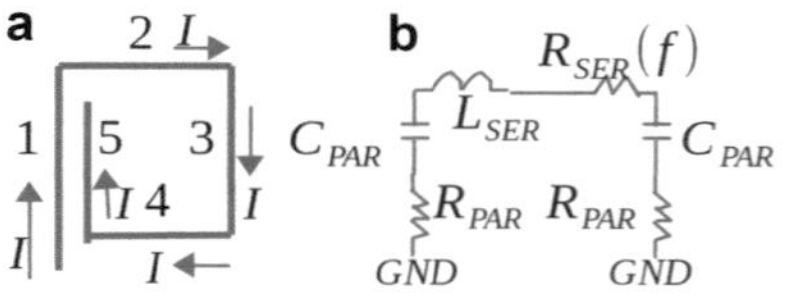

Fig. 2.8 (a, b) Square planar spiral inductor with five segments and equivalent lumped element electrical circuit

With reference to Fig. 2.8a, the total inductance is

$$L_{TOTAL} = L_1 + L_2 + L_3 + L_4 + L_5 + 2L_{15,MUTUAL}$$
$$- 2(L_{13,MUTUA:} + L_{24,MUTUAL} + L_{35,MUTUAL}).$$

The total self-inductance of this planar spiral inductor is

$$L_{TOTAL,SELF} = \sum 0.0002 l_i \left(\ln \left(\frac{2l_i}{GMD} \right) + \frac{AMD}{l_i} + \frac{\mu T}{4} - 1,25 \right)$$

where AMD and GMD are respectively the arithmetic and geometric means. T is the frequency correction factor. The expression for the GMD is

$$\ln (GMD) = \ln (d) - \frac{1}{12 \left(\frac{d}{w} \right)^2} - \ln \left(\frac{1}{60 \left(\frac{d}{w} \right)^4} \right) \dots$$

where d is the distance between the segments and w the width of each segment.
The Q factor for each segment is given by

$$Q_i = \frac{GMD}{l_y} + \ln \left(\frac{l_i}{GMD} + \sqrt{1 + \left(\frac{l_i}{GMD} \right)^2} \right)^{-2} - \sqrt{1 + \left(\frac{l_i}{GMD} \right)^2},$$

and then the mutual inductance between planar spiral inductor segments I,j is
$M_{i,j} = 0.00021_i Q_j$ and $AMD = w + t$ where w is the width and t the thickness of each segment. Figure 2.8b shows the equivalent lumped element electrical circuit of the square planar spiral inductor of Fig. 2.8a.

C_{PAR} is the lumped element equivalent circuit capacitance responsible for oddl even mode as well as interline coupling capacitances between coupled lines of the planar spiral inductor. The planar spiral inductor segments are separated into two groups, for evenlodd coupled pair mutual capacitance and inductance calculations. The capacitance between planar spiral segments is found from the equivalent circuit (Fig. 2.9). In this figure, C_{EVEN}, C_{ODD}, $C_{INTERLINE}$ are the even, odd, and interline coupling capacitances per unit length, respectively. The interline coupling capacitance is

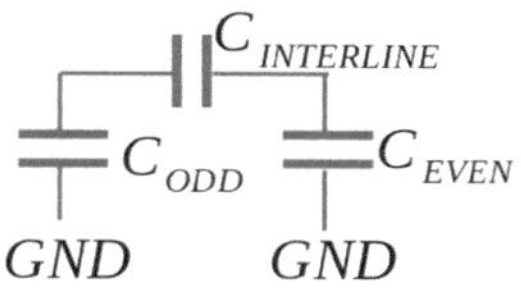

Fig. 2.9 Interline, even odd coupling between segments of planar spiral inductor

$$C_{INTERLINE} = \frac{C_{ODD} - C_{EVEN}}{2}$$

and the total capacitance is $C_{PAR} = 2C_{EQUIVALENT} + C_{ODD,OUTSIDE}\frac{F}{m}$

where the second term is the odd capacitance of the outside segment.

The frequency-dependent series resistance is $R(f) = \frac{2}{\omega}\sqrt{\frac{f\pi\mu_{CONDUCTOR}}{\sigma}}\frac{\Omega}{m}$ where $\mu_{CONDUCTOR}$, σ are respectively the magnetic permeability and conductivity of the conductor material.

The parasitic resistance is $R_{PAR} = \frac{\rho l}{hW}\frac{\Omega}{m}$

where h is the dielectric substrate thickness.

The quality factor (Q) of the inductor is $Q = \frac{3Z}{Z}$

where Z is the frequency-dependent resistance. The resonant frequency is

$$f_{RESO} = \frac{1}{2\pi\sqrt{C_{PAR}L_{TOTAL}}}\sqrt{\frac{1.0 - R_{SER}^2\frac{C_{PAR}}{L_{TOTAL}}}{1.0 - R_{PAR}^2\frac{C_{PAR}}{L_{TOTAL}}}}.$$

Of the four planar spiral inductor parameters, the series resistance is frequency dependent because of skin and proximity effects.

Skin effect is the result of an alternating electric current (AC) distributing itself within a conductor so that the current density (current per unit cross-sectional area of the conductor) is maximum at the surface of the conductor. The current density decreases exponentially with increasing depth in the conductor. The depth at which the current density becomes $\frac{1}{e}$ of its maximum value is called the skin depth.

The skin depth and effect are a result of Faraday's and Lenz's law and as embodied in Maxwell's equations for electromagnetic wave propagation in a conductor. Skin depth frequency dependent: as frequency increases, current flow gets more confined to regions of the conductor near the surface—current crowding. That is, the skin depth decreases, and the resistance increases. Skin effect can be tackled by using tube conductors, or braided wires.

The physics of skin effect is very interesting. A periodic, time-varying current I flowing through a conductor induces a periodic time-varying magnetic field $\vec{H}_{INDUCED}$. As the periodic time-varying current increases\decreases, a corresponding periodic time-varying, circulating eddy current I_{EDDY} which flows in a direction opposite to the input alternating current\voltage wave. This eddy current cancels most of the input current flow at the center of the conductor and reinforces it near the surface (Fig. 2.10). A close relation of the skin effect is the proximity effect (Fig. 2.11) conductors

Fig. 2.10 Skin effect in a conductor with periodic time-varying current\voltage input

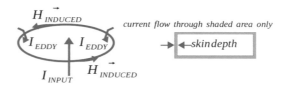

Fig. 2.11 Proximity effect between closely located parallel rectangular cross section

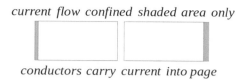

current flow confined shaded area only

conductors carry current into page

Fig. 2.12 Hexagonal and octagonal planar spiral inductor

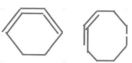

2.3.1   Manufacturability Constraints of Planar Spiral Inductors

Greenhouse [14] who extended the work of Grover [14] to derive the formulas due the planar spiral inductor used rectangular conductor segments to construct a square| rectangular planar spiral inductor. Rectangular conductor segments can be used to construct hexagonal or octagonal planar spiral inductors (Fig. 2.12).

However, the square|rectangular planar spiral inductor has emerged as the most popular design for a number of practical real-world constraints related to the manufacture of these inductors, arising from the fact that these planar spiral inductors were initially conceived of to fit inductors inside integrated circuits. As the available floor space inside an integrated circuit is very small, its use must be optimized. A square|rectangular planar spiral inductor satisfies this constraint best, compared to the other shapes (hexagon, octagon). In addition, all planar spiral inductors (especially those inside integrated circuits) are fabricated using the same techniques as used for manufacturing the semiconductor devices, i.e., a mask has to be constructed first, before the planar spiral indictor can be patterned on the integrated circuit floor. Once again, creating a mask for a square|rectangular planar spiral inductor is more easier or cost-effective than a mask for a hexagonal or octagonal planar spiral inductor. The most difficult mask to fabricate is that for a circular planar spiral indictor. The mask has to be made with tight tolerances, so that subsequent minor manufacturing issues would still not affect the correct layout of the planar spiral inductor.

Another key manufacturing constraint is that the planar spiral inductor must be located correctly so that it can be connected to the rest of the components inside the integrated circuit. In this case as well, the planar spiral inductor is the ideal choice to satisfy this criterion. Finally, each of the segments of a planar spiral inductor must be manufactured to maintain the correct interline spacing (within applicable tolerances), and the planar spiral inductor best satisfies this criterion, since its mask can be constructed most easily.

2.4   Current Sheet and Mohan Formalism for Planar Spiral Inductor

Electrical current confined to the surface of a conductor, without penetrating the volume, is a current sheet. It arises from the properties of fluids that conduct electric current, in which case, if there is an electric current through the volume, then internal magnetic forces expel the current from the volume, forcing the current to the surface.

An infinite current sheet can be visualized as an infinite planar array of parallel wires all carrying the same electric current I. Then, for N wires each carrying a current I per unit length, Ampere's law gives

$$\int \vec{B} \cdot \vec{\mathrm{dl}} = \mu_0 I_{enc}$$

where the integral is over a closed rectangle enclosing all the wires. For the sides of the rectangle orthogonal to the wires,

$$\vec{B} . \vec{\mathrm{dl}} = 0 \cos\left(\frac{\pi}{2}\right) = 0,$$

and for the two sides parallel to the plane of wires,

$$2 \int B \, \mathrm{dl} = \mu_0 I_{emc} \text{ or } B = \frac{\mu_0 I_{emc}}{2L} \text{ or } B = \frac{\mu_0 N I L}{2L} \text{ or } B = \frac{\mu_0 N I}{2}.$$

The largest current sheet in the solar system is the heliosphere current sheet, about 10,000 km thick, and extends from the Sun to beyond the orbit of the planet Pluto. Current sheets in plasmas store energy by increasing the energy density of the magnetic field. **The use of current sheets in planar spiral inductor analysis and design is justified because at 100 s of MHz–10s of GHz operating frequencies of integrated circuit embedded planar spiral inductors, skin and proximity effects force the current to flow in a thin layer of the conductor, at its surface.**

Contrary to the claims in [18], *the suggested technique fails to provide answer to the most important question that a planar spiral inductor designer faces.*

How many turns or segments of specified conductor width are required to fabricate a planar spiral inductor of specified target inductance value?

The current sheet-based approach is based on the previous work of Wheeler. The new expression for the inductance of a planar spiral inductor is

$L_{MOHAN} = \frac{n^2 \mu_0 d_{AVG} K_1}{1 + K_2 \rho}$ where $n. d_{AVG} \cdot K_1, K_2, \rho$ are the number of turns, the average diameter, layout-dependent parameters, and fill factor, respectively. The fill factor measures how "hollow" a planar spiral inductor is— $d_{OUTER} \simeq d_{INNER}$ the fill factor is small, whereas for $d_{OUTER} \gg d_{INNER}\rho$ is large. Thus, two planar spiral inductors with the same average diameters, but different fill ratios, will have different total inductances. This analysis can be expanded by replacing the planar spiral inductor segments with symmetrical current sheets, with equivalent current densities. For a square planar spiral inductor, four identical current sheets can be identified. The current sheets on opposite sides are parallel to each other, while the adjacent ones are orthogonal. Orthogonal current sheets have zero mutual inductance. Combining this with symmetry arguments, the calculation of the total inductance of a planar spiral inductor is simplified to computation of the inductance of one sheet and the mutual inductance between opposite current sheets, exploiting geometric properties as geometric mean distance (GMD), arithmetic mean distance (AMD), and arithmetic mean square distance (AMSD). The resulting expression is

$$L_{MOHAN} = \frac{C_1 d_{AVG} \mu n^2}{2} \left(\ln\left(\frac{C_2}{\rho}\right) + C_3 \rho + C_4 \rho^2 \right)$$

where $C_1, \ldots C_4$ are layout-dependent parameters, essentially curve fitting parameters. The accuracy of this expression degrades as the ratio of intersegment spacing to the segment width $\frac{s}{w}$ increases. In real world integrated circuit planar spiral inductors, $S \leqslant W$. It must be noted that planar spiral inductor with small intersegment spacing improves the magnetic coupling between them, and reduces total area occupied by the spiral. A large spacing reduces the intersegment capacitance.

Using detailed mathematical analysis and curve fitting techniques, the previous expression for the planar spiral inductor is modified to a monomial expression $L_{MOHAN,MONO} = \beta d_{OUT}^{\alpha_1} W^{\alpha_2} d_{AVG}^{\alpha_3} n^{\alpha_4} s^{\alpha_5}$ where $\alpha_1, \ldots \alpha_5$ are layout-dependent parameters. This equation is a monomial in the variables n, s, etc.

While Greenhouse [14] in his original formulation of the planar spiral inductor considered only the printed circuit board-based version, Mohan [18] has examined the planar spiral inductor fabricated onside an integrated circuit. The values of the various parameters used by Wheeler and Mohan are listed in Tables 2.1, 2.2 and 2.3. Figure 2.13a, b contains the planar spiral inductor configurations used by Mohan.

Table 2.1 Modified Wheeler planar spiral inductor parameters

Type	K_1	K_2
Square	2.34	2.75
Hexagon	2.33	2.83
Octagon	2.25	2.55

Table 2.2 Coefficients for current sheet scheme for planar spiral inductor

Type	C_1	C_2	C_3	C_4
Square	1.27	2/07	0.18	0.13
Hexagon	1.09	2.23	0.00	0.17
Octagon	1.07	2.29	0.00	0.19
Circle	1.00	2.46	0.00	0.2

Table 2.3 Curve fitting monomial coefficients

Type	β	$\alpha_1(d_{OUT})$	$\alpha_2(w)$	$\alpha_3(d_{AVG})$	$\alpha_4(n)$	$\alpha_5(s)$
Square	1.62E-3	−1.21	−0.147	2.4	1.78	−0.03
Hexagon	1.28E-3	−1.24	−0.174	2.47	1,77	−0.049
Octagon	1.33E-3	−1.21	−0.163	2.43	1.75	−0.049

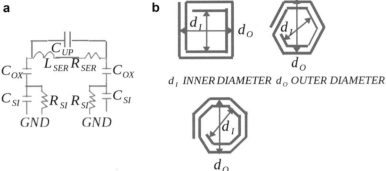

Fig. 2.13 (**a**) Equivalent lumped element circuit of planar spiral inductor fabricated inside an integrated circuitry. (**b**) Common planar spiral inductor structures

Fig. 2.14 (**a**) Equivalent lumped element electrical circuit for on-chip planar spiral inductor

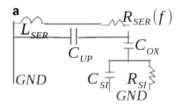

2.4.1 Planar Spiral Inductors Embedded Inside Integrated Circuits

The lumped element model of a planar spiral inductor fabricated on a silicon substrate inside an integrated circuit is shown in Figs. 2.13a and 2.14a. With reference to Fig. 2.14a, C_{SER}, L_{SER}, $R(f)_{SER}$ represent the series capacitance and inductance- and frequency-dependent series resistance, respectively. The series inductance, almost always computed with Greenhouse formalism, is the combination of the self-inductances of all the segments and the even/odd mutual inductance combinations of the parallel planar spiral inductor segments. The series capacitance arises from the underpass capacitance, i.e., one end of the planar spiral inductor has to be connected to the external circuit with a conductor strip placed below the planar spiral inductor, with an intervening dielectric layer, giving rise to this capacitor. The frequency-dependent resistance arises as a result of the skin and proximity effects.

C_{OX} is the capacitance due to the dielectric silicon dioxide layer immediately below the planar spiral inductor. The parallel combination C_{SI}, R_{SI} represents the parasitic capacitances that arise from the silicon substrate. The expressions for these planar spiral inductor parameters, for a total of N segments, are

$$C_{SER} = \frac{\epsilon_{OX} n w^2}{t_{OX,M1M2}}, C_{OX} = \frac{1 w \epsilon_{OX}}{2 t_{OX}} C_{SI} = \frac{lw C_{SUBSTRATE}}{2}$$

$$R_{SUBSTRATE} = \frac{2/W}{G_{SUBSTRATE}} R_{SER} = \frac{\rho l}{\delta w \left(1 - e^{\frac{-t}{\delta}}\right)}$$

where $n = N - 1$ is the number of overlaps of the underpass conductor and the planar spiral inductor and l, w, and t are respectively the total length, the width, and the thickness of each segment of the planar spiral inductor. t_{OX}, $t_{IX, M1M2}$ are respectively the oxide thickness and oxide thickness between the planar spiral inductor and the underpass. ϵ_{OX}, t_{OX} are respectively the oxide dielectric constant and oxide thickness. ρ, δ are respectively the substrate resistivity and skin depth of the conductor at the operating frequency. $C_{SUBSTRATE}$, $G_{SUBSTRATE}$ are respectively the capacitance and conductance per unit length, arising from transmission line theory concepts.

The key performance metric of an inductor is its quality factor Q, which is frequency dependent, and is degraded by the parasitic elements of the inductor.

The quality factor is defined as

$$Q = \frac{2\pi\{ \text{ Peak Magnetic Energy} - \text{Peak Electric Energy } \}}{\text{Energy Loss per Oscillation}}$$

$$E_{Electric,Peak} = \frac{V_O^2 C_O}{2}, E_{Magnetic,Peak} = \frac{L_{SER} V_O^2}{2\left(R_{SER}^2 + (L_{SER}\omega)^2\right)} \quad \text{and}$$

$$E_{LOSS} = \frac{\pi V_O^2}{2} \left(\frac{1}{R_{PARASTIC}} + \frac{R_{SER}}{R_{SER}^2 + (L_{SER}\omega)^2} \right)$$

2.4.2 *Frequency-Dependent Characteristics of Planar Spiral Inductors Embedded Inside Integrated Circuits*

In the lumped element equivalent circuit model of a planar spiral inductor, the series resistance is frequency dependent as a result of both skin and proximity effects. This means that at RF|microwave frequencies (199 s of MHz–10s of GHz), the key performance metric, the quality factor Q, will degrade as the frequency increases. With reference to Fig. 2.13a, the capacitor, inductor, and resistor triplet ((C_{UP}, $L_{SER}R_{SER}$(f)) and the capacitor and resistor triplet (C_{Ox}, C_{SI}, R_{SI}) form a Π network.

In this network, all components, except the frequency-dependent series resistance, are geometry dependent, so that in frequency-dependent analysis, only the resistor controls the behavior of the planar spiral inductor. Using the conventional notation for admittance

$(Y_{11}, Y_{12}, Y_{21}, Y_{22})$, the frequency-dependent inductance value of the planar spiral

inductor is $L_{SPIRAL,FREQ} = \dfrac{\Im\left(Y_{11}^{PRIME}\right)}{2\pi f}$ and $Q = \dfrac{\Im\left(\frac{1}{Y_{11}^{PRIME}}\right)}{\Re\left(\frac{1}{Y_{11}^{PRIME}}\right)}$.

The series admittance of the planar spiral inductor is defined as

$$Y_{SER}^{PRIME} = \frac{R_{SER}(f) + j\omega\left(C_{UP}(R_{SER}(f))^2 - L_{SER}(1 - \omega^2 C_{SER}L_{SER})\right)}{(R_{SER}(f))^2 + \omega^2 C_{UP}L_{SER}}$$

and the admittance of the parasitic oxide, substrate capacitors, and resistor is

$$Y_1^{PRIME} = \frac{(\omega C_{OX})^2 R_{SI} + j\omega C_{OX}\left(1 + (\omega R_S)^2 C_{SI}(C_{SI} + C_{OX})\right)}{1 + (\omega R_S(C_{SI} + C_{OX}))^2}.$$

The admittance for the other leg of the pi network is computed in a similar fashion.

Then, combining all these expressions, the input admittance of the planar spiral inductor is

$$Y_{11}^{PRIME} = Y_{SER}^{PRIME} + Y_1^{PRIME}.$$

This expression can then be optimized using available numerical analysis tools.

2.4.3 Mutual Capacitances and Inductances of Planar Spiral Inductors Embedded Inside Integrated Circuits

The two physical models of planar spiral inductors, in terms of microstrip transmissions lines, are the strip line and the microstrip line (Fig. 2.15a, b). While the microstrip model is more appropriate for planar spiral inductors with one side exposed to air, the strip line is more appropriate for planar spiral inductors embedded inside an integrated circuit or printed circuit board.

For both microstrip and strip lines, the mutual capacitance and inductance are given by

$$C_{MUTUAL} = \frac{C_{ODD}(\epsilon_r) - C_{EVEN}(\epsilon_r)}{2}$$

and the mutual inductance is

$$M = \frac{\epsilon_0 \mu_0}{2} \left(\frac{1}{C_{EVEN}(\epsilon_r = 1)} - \frac{1}{C_{ODD}(\epsilon_r = 1)} \right)$$

where ϵ_e is the relative dielectric constant.

For coupled microstrip lines, the even capacitance per unit length is

$$C_{EVEN}(\epsilon_r) = C_P + C_1 + C_{1^{PRIME}} \quad \text{where} \quad C_p = \frac{\epsilon_r \epsilon_0 W}{h},$$

w, h are trace width and dielectric thickness, respectively:

$$C_1 = \frac{1}{2} \left[\left[\frac{\sqrt{\epsilon_r}}{CZ_9} - C_p \right] \right] \quad \text{and} \quad C_{I^{PRME}} = \frac{C_1 \sqrt[4]{\frac{\epsilon_0}{\epsilon_e}}}{1 + \frac{Ah}{s} \tanh\left(\frac{10s}{h}\right)}$$

where s is the trace separation.

$A = e^{-0.1e^{23 - \frac{1-14}{h}}}$ and the characteristic impedance is given by

$$Z_0 = \frac{60}{\sqrt{\epsilon_0}} \ln\left(\frac{8h}{w} + \frac{w}{4h} \right) \frac{w}{h} \leqslant 1 \quad \text{and}$$

$$Z_0 = \frac{120\pi}{\sqrt{\epsilon_0}\left(1.393 + \frac{w}{h} + 0.667 \ln\left(1.444 + \frac{w}{h}\right)\right)} \frac{w}{h} > 1$$

$$\epsilon_e = \frac{\epsilon_r + 1}{2} + \frac{\epsilon_r - 1}{2} \sqrt{1 + \frac{10h}{w}}.$$

In this case, C_p, C_1, $C_{1\ PRIME}$ are the parallel and fringing capacitances, respectively.

For the odd propagation mode case,

$$C_{ODD}(\epsilon_r) = \frac{C_{OS}}{2} + C_{CPS}$$

where $C_{CPS} = \epsilon_0 \frac{K(K^{PRIME})}{K(K)}$ and $C_{OS} = 4\epsilon_0 \epsilon_r \frac{K(k_0)}{K(k_0^{PRIME})}$ and

$$k = \frac{s}{s + 2w}, \quad K^{PRIME} = \sqrt{1 - k_0^2} \quad \text{and}$$

$$k_0 = \tanh\left(\frac{\pi W}{4h}\right)\coth\left(\frac{\pi}{4}\left(\frac{w+S}{h}\right)\right), k_0^{PRIME} = \sqrt{1 - K_0^2}$$

In this case, $K(k)$, $K(k^{PRIME})$ are complete elliptic functions, defined as

$$\frac{K\left(k^{PRIME}\right)}{K(k)} = \frac{\ln\left(\frac{2\left(1+\sqrt{k^{PRIME}}\right)}{1-\sqrt{k^{PRIME}}}\right)}{\pi} \quad 0 \leq k < 0.707$$

$$\frac{K\left(k^{PRIME}\right)}{K(k)} = \frac{\pi}{\ln\left(\frac{2\left(1+\sqrt{k}\right)}{1-\sqrt{k}}\right)} \quad 0.707 \leq k < 1$$

A strip line mimics a planar spiral inductor embedded inside an integrated circuit, as it has the same dielectric material above and below the spiral. In this case,

$$C_{EVEN} = \frac{\sqrt{\epsilon_r}}{CZ_{EVEN}} C_{ODD} = \frac{\sqrt{\epsilon_r}}{CZ_{ODD}} \quad \text{and}$$

$$Z_{EVEN} = \frac{30\pi}{\sqrt{\epsilon_r}} \frac{K\left(k_{EVEN}^{PRIME}\right)}{K(k_{EVEN})} k_{EVEN}^{PRIME} = \sqrt{1 - K_{EVEN}^2} \quad \text{and}$$

$$Z_{ODD} = \frac{30\pi}{\sqrt{\epsilon_r}} \frac{K\left(k_{ODD}^{PRIME}\right)}{K(k_{ODD})} K_{ODD}^{PRIME} = \sqrt{1 - K_{ODD}^2}$$

Figure 2.15a, b and c illustrates these configurations.

2.5 Planar Spiral Inductor Transformers Embedded Inside Integrated Circuits

Armed with the concepts of mutual capacitance and inductance between parallel segments of a planar spiral inductor, planar spiral inductor-based transformers embedded inside integrated circuits can now be examined. Miniaturized microwave integrated circuits (MMIC) operating at 100 s of MHz or 10s of GHz require that passive circuit components, e.g., inductors, be embedded inside the integrated circuit, to ensure proper operation (e.g., minimize parasitic capacitances and inductances), reduce production costs, and reduce device size. Including inductors inside an integrated circuit is possible by using the same technology used to fabricate integrated circuit transistors. Therefore, by appropriately configuring planar spiral inductors inside an integrated circuit, an embedded transformer can be fabricated inside an integrated circuit.

Conventionally, for CMOS (complementary metal-oxide-semiconductor) device fabrication processes, transistors are placed on a substrate (e.g., silicon, gallium

Fig. 2.15 (**a**) Microstrip and strip line configurations. (**b**) Types of microstrip parasitic capacitances. (**c**) Evenǀodd mode microstrip parasitic capacitances

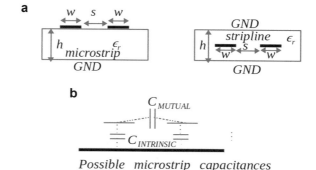

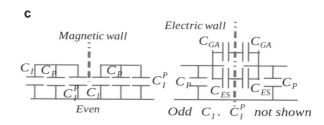

Possible microstrip capacitances

arsenide). Embedded(on-chip) transformers consisting of coplanaeǀvertically stacked planar spiral inductors are inefficient. This is unavoidable, since by their construction parasitic capacitative energy loss and magnetic flux leakage loss cannot be prevented. However these embedded planar spiral inductor transformers are very popular as they can be fabricated easily with existing integrated circuit(chip) manufacturing technology. As it is very difficult to integrate three-dimensional structures (e.g., solenoid) to standard CMOS process technologies, monolithic transformers are fabricated as interleaved planar spiral inductors. The planar geometry prevents full confinement of the magnetic field within the transformer, and the magnetic field passes through the substrate with finite resistivity resulting in magnetic losses, along with Ohmic losses, due to the thin metallization and parasitic capacitive losses between the conducting material of the planar spiral inductor transformer and substrate, and

also in between the segments of the planar spiral inductor transformer.

Despite being low efficiency devices, planar spiral inductor transformers are preferred as they are easy to fabricate (simple planar spiral rectangleǀsquare geometry and compatibility with CMOS fabrication processes). To achieve better performance, a number of novel variations of conventional planar spiral inductor configuration have been experimented with, e.g., stacked-type structures along with thick vias to connect multiple vertically separated metal layers, as well as air gap separation. In addition, MEMS (microelectromechanical systems) device fabrication procedures have also been adapted to embedded planar spiral inductor transformer fabrication.

Standard performance characteristic figures of merit must be defined to compare different planar spiral inductor transformer designs and evaluate their performance.

As a transformer consists of two coupled inductors, fundamental performance parameters of planar spiral inductor transformers are also similar to
those of inductors:

- Quality factor Q
- Self-resonance frequency f_{SR}
- Coupling factor (k)
- Mutual inductance (M)
- Insertion loss, power gain (G)
- Transformer's characteristic resistance (TCR)

These factors are interrelated, which makes design optimization difficult. For example, with a fixed number of turns, designing a transformer with a large footprint (substrate area) allows a larger width and cross-sectional area for the metal conductors. Increasing cross-sectional area for the conductors reduces Ohmic heating losses and improves the Q factor. Unfortunately, cross-sectional areas' related parasitic capacitances increase (large overlap area) with the substrate, and decrease with bandwidth. To achieve high Q factor for planar spiral inductor transformers, the self-resonance frequency would degrade. Therefore, the design\implementation of monolithic transformers depends on the intended application and related performance parameters.

Both the primary and secondary inductors of a planar spiral inductor transformer have their own Q factors, Q_P, Q_S, and together determine the overall performance of the planar spiral inductor transformer. The quality factor of an inductor indicates all the losses and the efficiency of that inductor. Therefore, the foremost goal of planar spiral inductor transformer design process is to maximize the Q factor for both the inductors. The Q factor is defined as

$$Q = \frac{\omega L_{SECONDARY}}{R} \, (\, substrate \; loss \; factor \,)(\, self - resonanr \; factor \,).$$

Another definition is in terms of admittance parameters Y:

$$Q = \frac{\Im Y_{11}}{\Re Y_{11}}$$

The Q factor is the ratio of energy stored in an inductor and energy dissipated in its series resistance, and can be improved by decreasing the series resistance. At RF\ microwave frequencies, the skin effect and proximity effects increase the series resistance and so degrade the Q factor. In addition, substrate loss imposes an upper limit on the Q factor. Fabricating the transformer on a higher resistivity substrate, such as glass, or shielding the conductors from the substrate increases the Q factor.

The coupling coefficient (k) measures the magnitude of magnetic coupling between the two inductors of a transformer and so is a key performance metric.

Greater overlap area and tighter packing of the two inductors result in a larger coupling coefficient. The coupling coefficient, k, is

$$k = \frac{M}{\sqrt{L_{PRI} L_{SEC}}}$$

where M is the mutual inductance between the two inductors and L_{PRI}, L_{SEC} are the self-inductances. The self-inductances depend on the physical length of the inductors, so they remain constant if the length remains fixed Therefore, the coupling coefficient, k, is the measure of the mutual inductance between primary and secondary inductors of a transformer. Stacked transformers have higher coupling coefficients, due to their tight geometry. But planar interleaved transformers have lower k values. There is a trade-off between the voltage transformation ratio n and k. For applications requiring large voltage or impedance transformation ratios

$$\left(n = \frac{N_1}{N_2} = \frac{V_1}{V_2} \right),$$

there should be a large variance in the number of turns of the primary to the secondary inductor, which decreases the overlap area between them, and increases leakage flux, resulting in lower k values.

Being coupled inductors, transformers exhibit resonance at particular frequencies called the self-resonance frequency of the transformer. One of the main design criteria for RF-CMOS embedded planar spiral inductor transformers is to make this resonance frequency high to increase the transformer's operating bandwidth. At resonance the transformer sucks in all input signal energy, and output is severely degraded. Transformers resonate at a particular frequency, due to the presence of parasitic capacitances:

- Between the substrate and the inductor, directly proportional to the area of the inductor facing the substrate
- In between the two inductors of the transformer

Beyond the resonant frequency, capacitive effects dominate over the inductive effects. The self-resonant frequency is

$$f_{SELF\ RESONANCE} = \frac{1}{2\pi \sqrt{L_{TOTAL} C_{TOTAL\ PARASITIC}}}.$$

An ideal transformer is lossless and delivers all input power without losses, at the output. Real-world transformers are lossy (i.e., insertion losses) which decrease the output power. So power gain (G) of a real-world transformer is a key performance metric, defined as the ratio of output to input power, defined as

$$G = 1 + 2\left(\alpha - \sqrt{\alpha^2 - \alpha} \right) \text{ where } \alpha = \frac{\Re(Z_{11})\Re(Z_{22}) - \{Z_{12}\}^2}{\{\Re(Z_{12})\}^2 + \{\Im(Z_{12})\}^2} \text{ and } Z_{11}, Z_{12} \cdot Z_{21}, Z_{22}$$

are the small signal impedances.

A transformer's characteristic resistance (TCR) is another performance metric of embedded planar spiral transformers, used to characterize these transformers when used as tuned loads. TCR combines Q and k parameters of a planar spiral transformer to evaluate its behavior. **When used as a tuned load in a circuit, the capacitive reactance of a transformer counterbalances the inductive reactance, allowing the circuit to be loaded purely by the resistive component.** This resistive component is also called as the input parallel resistance of a transformer and is a measure of the TCR, approximately doubling the value of the parallel input resistance. TCR is vital to designing the geometry and dimensions of a transformer to optimize the available output power and gain. TCR is expressed as

$$TCR = \left(\frac{\omega (kQ_p)^2 Q_S}{1 + k^2 Q_p Q_S} \right) \left(L_p \left(1 + \frac{1}{Q_p^2} + \frac{k^2 Q_S}{Q_p} \right) \right)$$

where k, L_p, Q_p, Q_S are respectively the coupling coefficient, primary inductance, and quality factors for the primary and secondary inductors.

Various lumped element electrical circuit and high-frequency circuit models for embedded planar spiral inductor transformers have been proposed as shown in Fig. 2.16a, b, c, d, and e. Each of these proposed models exploits mutual capacitance and inductance between two planar spiral inductors, to achieve the transformer

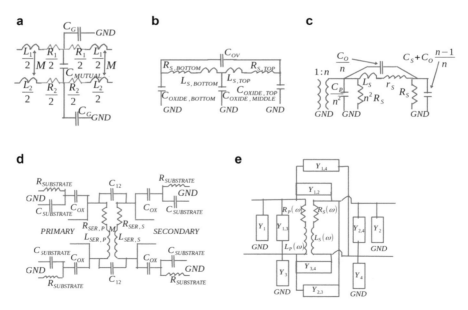

Fig. 2.16 (**a**) Lumped element equivalent circuit model for planar spiral inductor embedded transformers. (**b**) Vertically stacked planar spiral inductor embedded transformers. (**c**) Planar spiral inductor embedded transformers. (**d**) Planar spiral inductor embedded transformer. (**e**) High-frequency planar spiral inductor embedded transformer

effect. The high-frequency planar spiral inductor transformer model of Fig. 2.18e is based on the fact that that inductance at microwave frequencies is frequency dependent, so that the total impedance of the frequency-dependent series inductance and resistance combination of each planar spiral inductor is $Z(\omega) = R(\omega) + j\omega L(\omega)$.

Modeling of planar spiral inductor transformer during the design phase is essential to achieve the target transformer performance characteristics. This enables estimation of key performance metrics as the Q factor, mutual inductance, self-resonance frequency, and coupling coefficient prior to device fabrication. Embedded transformers are fabricated on silicon substrates, which draws in frequency-dependent parasitic capacitances and resistances (i.e., at high frequencies due to eddy currents) that are difficult to analyze|model accurately, and this makes the task challenging.

Equivalent electrical circuit models for planar spiral inductor transformer include lumped elements as well as distributed transmission line-based features, typically some combination of these two. The equivalent lumped element circuit model of Fig. 2.18b includes only inductor segment—substrate parasitic capacitances and interinductor segment parasitic capacitances only.

At moderate RF frequencies, frequency-dependent mutual resistance and inductance are often neglected, but cannot be ignored at high RF frequencies to accurately estimate embedded transformer performance metrics, as in Fig. 2.18d, e. The corresponding analytical methods have also been enhanced using concepts of geometric mean distance (GMD) to accurately estimate mutual inductance during embedded planar spiral inductor transformer design, as

$$M = \frac{\mu_0 l}{2\pi} \left(\ln \left(\frac{1}{d_m} + \sqrt{1 + \left(\frac{1}{d_m}\right)^2} \right) + \frac{d_m}{l} - \sqrt{1 + \left(\frac{1}{d_m}\right)^2} \right)$$

where d_m, l are respectively the distance between the two planar spiral inductors and l is the total length of overlap.

It has been found that using patterned ground shields significantly enhances embedded planar spiral inductor transformer performance characteristics as self-inductance, mutual inductance, coupling factor, and self-resonance while suppressing frequency-dependent parasitic capacitances.

. With the recent introduction of 5G cellular telephony systems (with carrier frequencies corresponding to wavelengths in the millimeter/submillimeter range) worldwide, embedded planar spiral inductor transformer analysis|design techniques have had to be modified to curtail the ultrahigh-frequency parasitic capacitances| inductances that degrade the performance characteristics of these transformers at these high frequencies. Also, narrower metal segment widths are used to minimize the substrate capacitive coupling effect, which shifts the self-resonance frequency of a transformer to higher values. Moreover smaller segment lengths, which improve the Q factor, by reducing the frequency-dependent series resistance, are used. A related issue is that conventional techniques for modeling embedded planar spiral

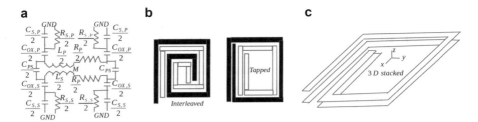

Fig. 2.17 (**a**) Improved ultrahigh-frequency embedded planar spiral inductor transformer lumped element model. (**b**) Embedded planar spiral inductor transformers—primary and secondary—are coplanar. (**c**) Embedded planar spiral vertically stacked inductor transformer

inductor transformers are valid for frequencies below 30 GHz. So, several attempts are being made to address this issue, e.g., accurately analyze skin effect in each segment of a planar spiral inductor, and then combine the results for adjacent nearest/ next nearest neighbor segments, using a combination of lumped circuit elements and analytical expressions. Experimentally obtained values for, e.g., self-resonance were compared with analytically calculated values, and appropriate modifications were made to both experimental techniques and the analytical methods, so that results agreed within applicable predefined tolerances. The improvements are included in the lumped element model in Fig. 2.17a.

Two simple layout schemes are used for embedded planar spiral inductor transformer. **The interleaved layout segments of one planar spiral inductor are fabricated in between the segments of the second planar spiral inductor, on the same plane,** thereby increasing the coupling coefficient of the two spiral inductors of the transformer. The total area occupied by the two planar spiral inductors is large, since segments of the second planar spiral inductor have to be placed in between the segments of the first planar spiral inductor, and each segment (whether it belongs to the first planar spiral inductor or the second) must have sufficient space in between itself and its two nearest adjacent neighbors, to maintain proper electrical isolation. Interleaved transformers are designed in 1:1 primary-secondary turn ratio. In 1:1 turn ratio configuration, the primary and secondary planar spiral inductors have matching lengths. A 1:N or N:1 turn ratio transformers require the planar spiral inductors to have different lengths, leading to magnetic flux leakage and signal energy waste.

The N turn inductor generates far more magnetic flux than the single turn inductor can accept.

The stacked layout is achieved by placing multiple metal layers on top of each other to make three-dimensional out-of-plane structures which ensures improved coupling. **Tight arrangement of planar spiral inductors with each other allows the magnetic flux to be confined within the structure and minimize flux leakage.** On the other hand, high parasitics, due to large overlap between vertically stacked metal segments, translates to low self-resonance frequency, and relatively difficult fabrication processes (each planar spiral inductor must be aligned within tight tolerances vertically). Because of their high coupling efficiency and small area, this type of embedded planar spiral inductor transformer is popular in RF-IC applications, and a lot of work has been done to maximize their performance characteristics. These configurations and their variants are shown in Fig. 2.17b, c.

2.5.1 Planar Spiral Inductor Transformer: Primary-Secondary Energy Transfer—Coupled Resonator Model

The simple twin LC tank circuit of Fig. 2.18a illustrates the key mechanism behind energy transfer between the primary and secondary planar spiral inductors of an embedded transformer. The circuit oscillates when the input impedance Z_{INPYT} is maximum. To determine the input impedance, the equivalent circuit of Fig. 2.18b is used. The impedance seen at port II is

$$Z_{\parallel} = s(L_1 - M) + \frac{1}{s(L_2 - M) + \frac{1}{sC_2}} + \frac{1}{sM} \quad \text{where}$$

$$s = j\omega \text{ and } L_1, L_2, M$$

are respectively the self and mutual inductances of the primary and secondary planar spiral inductors. Then

$$\frac{1}{Z_{INPUT}} = \frac{1}{Z_{\parallel}} + SC_1.$$

The condition for resonance is $1 + (C_1 C_2 L_1 L_2 - M)s^4 + (C_1 L_1 + C_2 L_2)s^2 = 0$. This is a $C = C_1 = C_2$ biquadratic equation in s, with four roots, of which the two negative ones can be discarded (because negative frequencies do not have any real-world physical meaning). The positive roots are

$$\omega_1 = \sqrt{\frac{-(C_1 L_1 + C_2 L_2) + \sqrt{C_1 L_1 + C_2 L_2)^2 + 4C_1 C_2 (M^2 - L_1 L_2)}}{C_1 C_2 (M^2 - L_1 L_2)}}$$

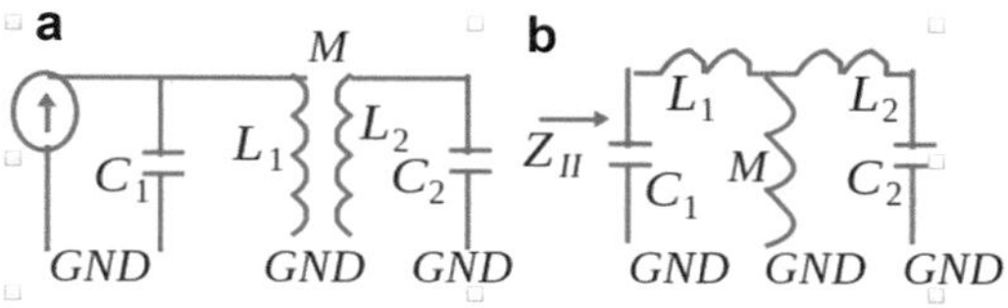

Fig. 2.18 (a, b) Coupled resonators and equivalent circuit got input impedance calculation

$$\omega_2 = \sqrt{\frac{-(C_1L_1 + C_2L_2) - \sqrt{(C_1L_1 + C_2L_2)^2 + 4C_1C_2(M^2 - L_1L_2)}}{C_1C_2(M^2 - L_1L_2)}}$$

Under resonance conditions $C = C_1 = C_2$ and $L = L_1 = L_2$, and then the positive resonance frequencies can be reduced to

$$\omega_1 = \frac{1}{\sqrt{C(L + M)}}, \quad \omega_2 = \frac{1}{\sqrt{C(L - M)}}, \quad \text{and} \quad Q = \frac{1}{R}\sqrt{\frac{L}{C}}.$$

2.5.2 Planar Spiral Inductor Transformer: Concentrating Magnetic Flux with Ferrite Layer

In order to concentrate and confine the magnetic flux to the three-dimensional region occupied by a stacked planar spiral inductor transformer, a recent enhancement to the basic stacked planar spiral inductor embedded transformer is to have a high resistivity ferrite layer right after the oxide layer. The equivalent lumped element electrical for a single planar spiral inductor is shown in Fig. 2.19a, b. And the corresponding equivalent lumped element embedded planar spiral inductor with two of these inductors stacked vertically is shown in Fig. 2.19c The Ohmic resistance of the two planar spiral inductors are
$R_{CU,P} = \frac{\rho_{Cu}I_{T,P}}{t_{M,P}W_P}$ and $R_{CU,S} = \frac{\rho_{Cu}I_{T,S}}{t_{M,S}W_S}$ where $I_{T,P}$, $I_{T,S}$, $t_{M,P}$, $t_{M,S}$, W_P, W_S are respectively the total length, thickness, and widths of the primary and secondary planar spiral inductors. The resistances of the two magnetic layers are

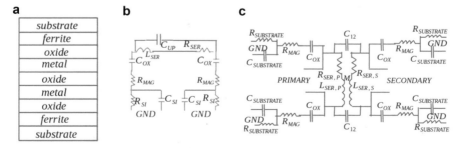

Fig. 2.19 (a) Cross section of embedded planar spiral inductor transformer with ferrite layer below oxide layer. (b) Equivalent lumped element electrical circuit of standalone planar spiral inductor with ferrite layer below oxide layer. (c) Embedded planar spiral inductor transformer with ferrite layer below oxide layer for each planar spiral inductor

$$R_{MAG,P} = \frac{2\rho_{MAG}d_{MAG}}{l_{T,P}W_P} \quad \text{and} \quad R_{MAG,S} = \frac{2\rho_{MAG}d_{MAG}}{l_{T,S}W_S}.$$

The resistances of the silicon substrates are

$$R_{Si,p} = \frac{2\rho_{Si}d_{Si}}{l_{T,p}W_p} \quad \text{and} \quad R_{Si,S} = \frac{2\rho_{Si}d_{Si}}{l_{T,S}W_S}.$$

In a similar fashion, the capacitances of the silicon substrate for the primary and secondary planar spiral inductors are

$$C_{SUBSTRATE,P} = \frac{\epsilon_0\epsilon_{r,Si}l_{T,P}W_P}{2d_{Si}} \quad \text{and} \quad C_{SUBSTRATE,S} = \frac{\epsilon_0\epsilon_{r,Si}l_{T,S}W_S}{2d_{Si}},$$ and the oxide capacitances are

$$C_{OX,P} = \frac{\epsilon_0\epsilon_{r,ox}l_{T,P}W_P}{2t_{ox}} \quad \text{and} \quad C_{OX,S} = \frac{\epsilon_0\epsilon_{r,ox}l_{T,S}W_S}{2t_{ox}}$$ where d_{Si}, d_{MAG}, t_{OX} are respectively the physical lengths of the silicon, oxide, and oxide layers. The coupling capacitance between the two planar spiral inductors is

$$C_{COUP} = \frac{\epsilon_0\epsilon_{r,ox}d_{OUT}^2}{t_{OX}}.$$

The total inductance of each of the planar spiral inductors can be calculated by either the Greenhouse scheme or the current sheet method.

The scattering parameters, very useful for evaluating small signal performance characteristics of the embedded planar spiral inductor transformer, can be evaluated in terms of the ABCD parameters for each subsection, and then cascaded together. The components of the primary planar spiral inductor are A = 1, B = 0,

$$C = \frac{1}{R_{MAGNEYIC,P} + \frac{1}{j\omega C_{OX,P}} + \frac{R_{SUBSTRATE,P}}{1+j\omega C_{SUBSTRATE,P}R_{SUBSTRATE,P}}}, \quad \text{and} \quad D = 1$$

Similarly, for the secondary planar spiral inductor of the transformer, A = 1, B = 0

$$C = \frac{1}{R_{MAGNEYIC,S} + \frac{1}{j\omega C_{OX,S}} + \frac{R_{SUBSTRATE,S}}{1+j\omega C_{SUBSTRATE,S}R_{SUBSTRATE,S}}}, \quad \text{and} \quad D = 1.$$

Similarly, the components of the magnetic coupling ABCD matrix are

$$A = \frac{-L_P}{M} \quad B = \frac{-j\omega L_p L_s}{M} + j\omega M \quad C = \frac{-1}{j\omega M}, \quad \text{and} \quad D = \frac{-L_S}{M}$$

where $L_P, L_S, M = k\sqrt{L_P L_s}, \omega$ are respectively the primary and secondary self-inductances and M the mutual inductance.

Denoting A_F, B_F, C_F, D_F as the elements of the final cascaded ABCD matrix, the S parameters can be expressed as

$$S_{11} = \frac{A_F + \frac{B_F}{Z_0} - C_F Z_0 - D_F}{A_F + \frac{B_F}{Z_0} + C_F Z_0 + D_F} \quad S_{12} = \frac{2(A_F B_F - C_F D_F)}{A_F + \frac{B_F}{Z_0} + C_F Z_0 + D_F}$$

$$S_{21} = \frac{2}{A_F + \frac{B_F}{Z_0} + C_F Z_0 + D_F}$$

and

$$S_{22} = \frac{-A_F + \frac{B_F}{Z_0} - C_F Z_0 + D_F}{A_F + \frac{B_F}{Z_0} + C_F Z_0 + D_F}.$$

Then using the familiar notation, the impedance parameters are

$$Z_{11} = \frac{Z_0((1 + S_{11})(1 - S_{22}) + S_{12}S_{21})}{(1 - S_{11})(1 - S_{22}) - S_{12}S_{21}}, \quad Z_{12} = \frac{2Z_0 S_{12}}{(1 - S_{11})(1 - S_{22}) - S_{12}S_{21}}$$

$$Z_{21} = \frac{2Z_0 S_{21}}{(1 - S_{11})(1 - S_{22}) - S_{12}S_{21}} \quad \text{and finally}$$

$$Z_{22} = \frac{Z_0((1 - S_{11})(1 + S_{22}) + S_{12}S_{21})}{(1 - S_{11})(1 - S_{22}) - S_{12}S_{21}}.$$

Using these results, the primary and secondary planar spiral inductor inductances and resistances are given by

$$L_p = \frac{\Im(Z_{11_i})}{w}, L_s = \frac{\Im(Z_{22})}{w}, R_p = \Re(Z_{11}), R_s = \Re(Z_{22}).$$

The key performance metric of any inductor is the quality factor (Q), and as an embedded planar spiral inductor transformer consisting of two coupled planar spiral inductors, the Q factor is defined as $Q = \frac{2\pi \; stored \; energy}{dissipared \; energy}$.

The stored energy is the magnetic field energy, and the quality factor can be redefined in terms of the difference in magnetic field energy of the two planar spiral inductors:

$$Q = \frac{\omega L R_C}{R_S\left(R_C + R_S\left(1 + \left(\frac{\omega L}{R_S}\right)^2\right)\right)}\left(1 - \frac{R_S^2(C_C + C_S)}{L} - \omega^2 L(C_C + C_S)\right)$$

where C_C, R_C are the coupling capacitance and resistance, respectively,

C_s, R_s are the series capacitance and resistance and L the inductance. Both coupling capacitance and resistance are related to the oxide and substrate capacitances and resistances, respectively:

$$R_C = \frac{1}{R_{Si}\omega^2 C_{OX}^2} + \frac{R_{Si}(C_C + C_{OX})^2}{C_{OX}^2} \quad \text{and}$$

$$C_C = \frac{C_{OX}\left(1 + (\omega R_{Si})^2 (C_{OX} + C_{Si})\right)}{1 + (\omega R_{Si}(C_{OX} + C_{Si}))^2}.$$

In terms of the small signal impedance parameters, the quality factors of the primary and secondary planar spiral inductors are

$$Q_P = \frac{\Im(Z_{11})}{\Re(Z_{11})} \quad \text{and} \quad Q_S = \frac{\Im(Z_{22})}{\Re(Z_{22})}.$$

2.5.3   Novel Application of Embedded Planar Spiral Inductors: Magneto-Impedance Effect—Integrated Circuit Compasses

The magneto-impedance effect [43] is a direct consequence of the skin effect and arises when a high-frequency signal passes through an *amorphous* magnetic material wire, in the presence of an external magnetic field. As the impedance of the wire changes, it can be detected, and thereby correlated with the external magnetic field, thereby acting as a magnetic field detector. Exploiting the skin effect means the amorphous wire core magnetization is avoided, negating magnetic noises. This effect is utilized to fabricate practically sensitive linear micromagnetic sensor and electronic sensor circuits in which a zero-magnetostrictive amorphous wire with a pick-up inductor with a CMOS integrated multivibrator pulse voltage generator feeding the pulse current via a Schottky diode to the amorphous wire, eliminating the need for high-frequency signal generator.

Magneto-impedance is controlled by how deep a periodic time-varying current can penetrate inside a conductor. The penetration length is the skin depth, increases as the square root of the electrical resistivity of the material, and is inversely proportional to the square root of the product of the permeability and the frequency of the applied periodic time-varying current. So, in materials with high values of permeability, the skin depth can be much less than the thickness of the conductor even for moderate applied AC frequencies.

An external applied magnetic field decreases the permeability enabling greater penetration of the applied time-varying current into the bulk of the magnetic material, leading to clearly observable variations in-phase and out-of-phase components of the magneto-impedance. The magnitude of the applied magnetic fields can vary from the value of the magnitude of the Earth's magnetic field to few tens of Oersted.

The magnetic noise in the amorphous magnetic wire is Barkhausen noise caused by the random magnetic domain wall displacements (of the spike domains). The demagnetizing field

$$\vec{H}_{dem} = \frac{-N_{dem}\vec{M}}{\mu_0},$$

where $N_{dem}, \vec{M}$ are respectively the demagnetization coefficient and magnetization. This demagnetizing field appears at the edges of the core and reduces the sensitivity of the amorphous wire sensor. Therefore, the skin effect is key to realizing this magneto-impedance effect, as it forces current flow through a thin outer region of the amorphous wire.

The impedance of a magnetic wire is

$$Z = \frac{R_{DC}kaJ_0(ka)}{2J_1(ka)}$$

where J_0, J_1 are Bessel's functions and a the wire radius. The DC resistance is $R_{DC} = \frac{\rho l}{\pi a^2}$, and $k = (1-j) = \sqrt{\left(\frac{2\rho}{\mu\omega H_{EXT}}\right)}$ where μ is the maximum circumferential differential permeability and ρ the resistivity of the amorphous wire material. When the skin depth is much greater than the wire radius,

$$\omega \ll \frac{2\rho}{\mu a^2} \text{ and } Z = R_{DC} + \frac{j\mu\omega l}{8\pi}.$$

In the other case, when the skin depth is much smaller than the wire radius,

$$Z = \frac{(1+j)aR_{DC}}{2}\sqrt{\frac{\omega\mu H_{EXT}}{2\rho}}$$

which is the magneto-impedance effect. The sensitivity of this amorphous wire sensor is $\frac{\partial\{Z\}}{\partial H_{EXT}} = \frac{1}{4\pi a}\sqrt{\frac{\rho\omega}{\mu}}$.

Figure 2.20 shows the amorphous wire domain model.

As ω increases, $\mu = \mu_R + j\mu_1$. The minimum magnetic noise spectral density β for single domain magnetization rotation model in the amorphous wire magneto-impedance effect is

Fig. 2.20 Domain model for amorphous wire

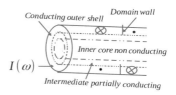

$$\beta = \sqrt{\frac{2\alpha k_B T}{\gamma \pi d / M_S}} \quad \text{where}$$

α, T, k_B, M_S, γ, d, l are respectively the magnetic damping constant, the temperature in Kelvin, Boltzmann's constant, saturation magnetization, gyromagnetic ratio, the wire diameter, and length.

Some existing uses of magneto-impedance are in electronic compasses for smartphone and wrist watches, and ultrasensitive (pico-Tesla) biomagnetic sensors.

2.6 Planar Antennas: Dipole, Loop, and Patch

Although being planar inductors, there is a key difference between planar inductors and planar antennas [44–52]. All antennas are operated in resonance, else they would not work. A receiver antenna would not be able to absorb all incoming signal energy unless it is operating at resonance. In line with this, an antenna must always be impedance matched to the rest of the receiving|transmitting circuitry.

The input impedance of a series lumped element RLC resonator is

$$Z_{INPUT} = R + j\omega L + \frac{1}{j\omega C},$$

and the input power delivered is

$$P_{INPUT} = \frac{\{I\}^2}{2}\left(R + j\omega L + \frac{1}{j\omega C}\right).$$

As expected, energy is dissipated in the resistor $P_{DISS,R} = \frac{\{I\}^2 R}{2}$. The energy stored in the electric and magnetic fields of the capacitor and inductor respectively are $P_C = \frac{\{I\}^2}{4\omega^2 C}$, $P_L = \frac{\{I\}^2 L}{4}$.

Then combining all these results, $P_{INPUT} = \frac{\{I\}^2}{2}\left(R + j\omega\left(\frac{L}{2} - \frac{1}{2\omega^2 C}\right)\right)$.

Now, the input impedance can be re-written as

$$Z_{INPUT} = \frac{2P_{INPUT}}{\{I\}^2} = \frac{2P_R}{\{I\}^2} = R,$$

since at resonance, average energy stored in the electric field of the capacitor and inductor is the same. Then, $L = \frac{1}{\omega_{RESO}^2 C}$ $\omega_{RESO} = \frac{1}{CL}$.

The quality factor Q of a resonator is defined as

$$Q = \frac{avg.\,energy\,stored}{energy\,loss/unit\,time} \quad \text{or}$$

$Q = \frac{\omega(E_{ELECTRIC\,FIELD} + E_{MAGNETIC\,FIELD})}{LOSS_{RESISTOR}}$ and is a measure of energy loss of the resonator. At resonance, the energy stored in the electric and magnetic fields are the same, so that

$$Q = \frac{\omega_{RESO}L}{R} = \frac{1}{\omega_{RESO}CR}.$$

Clearly, the Q factor varies inversely with the Ohmic resistance, and so the smaller the resistance, the higher the value of Q.

The input impedance can be expressed in terms of the resonant frequency and Q factor as

$$Z_{INPUT} = R + j\omega L\left(1 - \frac{1}{\omega^2 CL}\right) = R + j\omega L\left(\frac{\omega^2 - \omega_{RESO}^2}{\omega^2}\right).$$

This expression can be simplified as

$$Z_{INPUT} = R + \frac{2jQR\Delta\omega}{\omega_{RESO}}.$$

which can be simplified further in the special case the resonator is quasi-lossless. The bandwidth of the resonator is $BW = \frac{1}{Q}$.

For the parallel RLC resonator, the input impedance is

$$Z_{INPUT} = \left(\frac{1}{R} + \frac{1}{j\omega L} + j\omega C\right).$$

The corresponding input power is

$$P_{INPUT} = \frac{\{V\}^2}{2}\left(\frac{1}{R} + \frac{1}{j\omega L} + j\omega C\right).$$

The Ohmic energy loss is
$P_{LOSS} = \frac{\{V\}^2}{2R}$, and the energies stored in the capacitor's electric field and the inductor's magnetic field are

$$E_{ELECTRIC} = \frac{C\{V\}^2}{4} \quad \text{and} \quad E_{MAGNETIC} = \frac{\{V\}^2}{4\omega^2 L}.$$

Then the input power can be re-expressed as

$$P_{INPUT} = P_{LOSS} + 2j\omega(E_{MAGNETIC} - E_{ELECTRIC}),$$

and so the input impedance can be re-written as

$$Z_{INPUT} = \left(\frac{2}{\{1\}^2}\right) P_{LOSS} + 2j\omega(E_{MAGNETIC} - E_{ELECTRIC})$$

At resonance, the energies stored in the electric and magnetic fields are the same, so that the input impedance is simply the resistance. The resonance frequency is

$$\omega_{RESO} = \frac{1}{\sqrt{CL}},$$

and the quality factor can be expressed as

$$Q = \frac{R}{\omega_{RESO}L} = \omega_{RESO}CR :$$

and then the input impedance can be re-expressed as:

$$Z_{INPUT} = \frac{R}{1 + \frac{2jQ\Delta\omega}{\omega_{RESO}}}.$$

Real-world resonators are almost always connected to a load circuit, so the resonator Q value also varies depending on whether it is connected to a load. In case of a series resonator, the load resistance R_L gets added to the resonator resistance, and the overall Q value is reduced. In case of a parallel RLC resonator, the load resistor in combination with the resonator resistance increases the overall Q value. The situation is more complicated with complex loads. The external quality factor is

$$Q_{EXT} = \frac{\omega_{RESO}L}{R_L}$$

for series resistive loads, and

$$Q_{EXT} = \frac{R_L}{\omega_{RESO}L}$$

for parallel resistive loads. The loaded and unloaded Q values of the resonator are related as

$$\frac{1}{Q_L} = \frac{1}{Q_{EXT}} + \frac{1}{Q_{UNLOADED}}.$$

For most electronic circuits operating at 100 s of MHz to 10s of GHz, the dimensions of the circuit elements are comparable to wavelength of the input signal, so that lumped circuit components need to be replaced by transmission lines.

A transmission line of physical length l and characteristic impedance Z_0 is shorted at one end. The input impedance is $Z_{INPUT} = Z_0 \tanh (\alpha + j\beta)l$, which can be re-written as

$$Z_{INPUT} = \frac{Z_0(\tanh(\alpha l) + j\tan(\beta l))}{1 + j\tan(\beta l)\tanh(\alpha l)}.$$

Real-world transmission lines have small losses, so that a frequency can be written as L $\omega = \omega_{RESO} + \Delta\omega$,, so that for a pure TEM transmission line,

$$\beta l = \frac{\omega_{RESO}l}{V_{PHASE}} + \frac{\Delta\omega l}{V_{PHASE}}$$

As $I = \frac{\lambda}{2} = \frac{\pi V_{PHASE}}{\omega_{RESO}} \omega = \omega_{RESO} \; \beta 1 = \pi + \frac{\pi\Delta\omega}{\omega_{RESO}}.$
and so $\tan(\beta l) = \tan\left(\frac{\pi\Delta\omega}{\omega_{RESO}}\right)$

Consequently, the input impedance can be re-written as

$$Z_{INPUT} = Z_0\left(\alpha \mid + \frac{\pi\Delta\omega}{W_{RESO}}\right).$$

Now recognizing that the expression for the input impedance is of the form

$$Z_{INPUT} = Z_0(\alpha \mid + 2jL\Delta\omega)R = Z_0\alpha lL = \frac{\pi Z_0}{2\omega_{RESO}},$$

and the value for C can be obtained from resonance conditions. The quality factor is $Q = \frac{\beta}{2\alpha}$

Antiresonance (parallel type of resonance) can be obtained by using a shorted transmission line of length $\frac{\lambda}{4}$. The input impedance is

$$Z_{INPUT} = Z_0 \left(\frac{1 - j \tanh (\alpha l) \cot (\beta l)}{\tanh (\alpha l) - \cot (\beta l)} \right).$$

At resonance, $l = \frac{\lambda}{4}$ and any frequency $\omega = \omega_{RESO} + \Delta\omega$. Then, $\beta l = \frac{\pi}{2} + \frac{\pi\Delta\omega}{\omega_{RESO}}$, so that $\cot (\beta 1) = - \tan \left(\frac{\pi\Delta\omega}{\omega_{RESO}} \right)$.

Using these expressions, the input impedance can be re-written as

$$Z_{INPUT} = \frac{1}{\frac{1}{R} + 2j\Delta\omega C}$$

Then, $R = \frac{Z_0}{\alpha l}$, $C = \frac{n}{4\omega_{RESO}Z_0}$, and $Q = \frac{\beta}{2\alpha}$.

Both the precious two transmission line resonators involved transmission line lines short circuited at one end. Now a transmission line resonator with both ends open is examined. The input impedance is

$$Z_{INPUT} = Z_0 \left(\frac{l + j \tanh (\alpha l) \tan (\beta l)}{\tanh (\alpha l) + j \tan (\beta l)} \right).$$

At resonance, $l = \frac{\lambda}{2} \omega = \omega_{RESO} + \Delta\omega,$, so that

$$\beta l = \pi + \frac{\pi\Delta\omega}{\omega_{RESO}}.$$

Then,

$$\tan (\beta l) = \tan \left(\frac{\pi\Delta\omega}{\omega_{RESO}} \right),$$

and the input impedance becomes $Z_{INPUT} = \frac{Z_0}{\alpha l + \frac{j\pi\Delta\omega}{\omega_{RESO}}}$.

Then in a straightforward manner, $R = \frac{Z_0}{\alpha l}$, $C = \frac{\pi}{2\omega_0 Z_0}$, $Q = \frac{\beta}{2\alpha}$.

2.6.1 Planar Patch Antenna and Inverted F Antenna

Microstrip patch antennas [44–53] are low profile and are used where size, weight, cost, performance, and ease of installation are constraints that must be satisfied. They are used in all cellular phones, portable wireless communication equipment, as well as high-performance aircraft, spacecraft, satellite, and missile applications. These antennas are inexpensive to manufacture using modern printed circuit technology. They are very versatile in terms of resonant frequency, polarization, pattern, and impedance for a particular patch shape and mode. Even more interesting, by adding loads between the patch and ground plane (pins, varactor diodes), adaptive elements with variable resonant frequency, impedance polarization, and pattern can be designed. Operational disadvantages of microstrip antennas (low efficiency and power, high Q, poor polarization purity, poor scanning performance, spurious feed radiation, and very narrow frequency bandwidth) can be tackled with workarounds as increasing the height of the substrate to boost the efficiency and bandwidth. Unfortunately, surface waves are introduced which extract power from the input signal power. The surface waves travel within the substrate and they are scattered at bends and surface discontinuities and degrade the antenna pattern and polarization characteristics. Surface waves can be eliminated, while maintaining large bandwidths, by using cavities stacking, etc., can also be used to increase the bandwidth.

For real-world planar patch antennas, the conductor thickness and the substrate thickness must be much smaller than the free space wavelength of the signal be received/transmitted, i.e., $t \ll \lambda_0 \; 0.003\lambda_0 \leq h \leq 0.05\lambda_0$. Another requirement is that the patch antenna's pattern maximum is normal to the patch plane. The physical length of patch must be one third to half the free space wavelength. Given that the relative dielectric constants vary between 2.2 and 12, the ones that are most desirable for good antenna performance are thick substrates with dielectric constants near the lower range limit because they provide improved efficiency, broader bandwidth, and loosely bound fields for radiation into space. The disadvantage is larger element size. Thin substrates with higher dielectric constants are used for microwave circuitry to ensure tightly bound fields to minimize undesired radiation, coupling, and lead to smaller element sizes, but suffer greater losses and reduced efficiency and bandwidth. Since microstrip antennas are often integrated with other microwave circuitry, careful choice of trade-offs must be selected between good antenna performance and circuit design. Figure 2.21a, b, c shows a basic planar rectangular patch antenna and its lumped element equivalent electrical circuit.

The two radiating slots of the planar rectangular patch antenna demonstrate the fringing effects that are responsible for electromagnetic wave radiation. Similar fringing effects are also present along the two edges of the same patch antenna that do not have the radiation slots. Fringing is a function of the dimensions of the

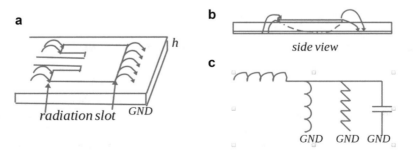

Fig. 2.21 (**a**) Planar rectangular patch antenna with two radiation slots. Arrows indicate direction of electric fields. (**b**, **c**) Planar rectangular patch antenna side view, and simple parallel RLC resonator lumped element equivalent circuit

patch and substrate thickness. For the principal E-plane (XY-plane), fringing is a function of $\frac{l}{h}$ and the relative dielectric constant ϵ_r of the substrate. Fringing influences the resonant frequency since both the length l and width are greater than the substrate thickness. The planar rectangular patch antenna supports quasi-TEM electromagnetic waves, since the relative dielectric constants of media on the two sides of the antenna are different. Of the fringing electric field lines, some are inside the dielectric substrate and some are partially in the air and partially inside the dielectric. Fringing makes the patch antenna appear wider electrically compared to its physical dimensions, as fringing occurs around the corners as well. Consequently, the relative dielectric constant is replaced by an effective relative dielectric constant given by

$$\epsilon_{r,EFF} = \frac{1+\epsilon_r}{2} + \frac{\epsilon_r - 1}{2}\left(\frac{1}{\sqrt{1 + \frac{12h}{W}}}\right).$$

Another interesting consequence of fringing is that the physical length appears increases by length extension, which, normalized to the substrate thickness, is

$$\frac{\Delta l}{h} = \frac{0.412(\epsilon_{r,EFF} + 0.3)\left(\frac{W}{h} + 0.264\right)}{(\epsilon_{r,EFF} - 0.258)\left(\frac{W}{h} + 0.8\right)}.$$

The length, for the dominant TM_{010}, mode is $l_{eff} = l + 2\Delta l$, and the corresponding resonant frequency is

$$f_{RESO,010} = \frac{1}{2l_{eff}\sqrt{\epsilon_{r,EFF}}\sqrt{\mu_0\epsilon_0}}.$$

Almost all antennas are operated in resonance conditions, and accurate impedance matching is essential to minimize signal loss. Figure 2.21c shows the equivalent transmission model for a twin radiation slot rectangular planar patch antenna. Here,

$$Y_1 = G_1 + jB_1 \text{ where } G_1 = \frac{W}{120\lambda_0}\left(1 - \frac{(hk_g)^2}{24.0}\right) \text{ and}$$

$$B_1 = \frac{W}{120\lambda_0}(1 - 0.636\ln(k_0 h))\frac{h}{\lambda_0} < \frac{1}{10} \text{ , and}$$

$$Y_1 = Y_2 \quad B_1 = B_2 \quad G_1 = G_2.$$

At resonance, $Y_2^{PRIME} = G_2^{PRIME} + jB_2^{PRIME} = G_1 - jB_1$, so that the input admittance and impedance can be written as

$$Y_{INPUT} = Y_1 + Y_2^{PRIME} = 2G_1 \ \ Z_{INPUT} = R_{INPUT} = \frac{1}{2G_1}.$$

To account for mutual interaction between the radiating slots, the expression for the resonant input impedance is re-written as

$$R_{INPUT} = \frac{1}{G_1 \pm G_{12}}$$

where the positive sign refers to the antisymmetric mutual interaction, and the negative sign refers to the symmetric mutual interaction. Antisymmetric mutual interaction occurs when the currents/voltages on the two slots are of opposite polarities. The general expression for G_{12} is

$$G_{12} = \frac{1}{\{V_0\}} \Re \left[\iint \overrightarrow{E_1} \times \overrightarrow{H_2}^{PRIME} ds \right]$$

where $\overrightarrow{E_1}, \overrightarrow{H_2}, V_0$ are respectively the electric field at slot 1, magnetic field at slot 2, and voltage across the slots. This expression can be re-written in terms of Bessel's functions as

$$G_{12} = \frac{1}{120\pi^2} \int \left[\frac{\sin \left(\frac{k_0 W}{2} \cos (\theta) \right)^2}{\cos (\theta)} \right] J_0(k_0/\sin (\theta))(\sin (\theta))^3 d\theta.$$

where $0 \leq \theta \leq \pi$.

In case the mutual interaction between the two radiating slots is neglected, the input resistance at an inset point X_0 is

$$R_{INPUT}(x = x_0) = R_{INPUT}(x = 0)\left(\cos \left(\frac{\pi x_0}{1} \right) \right)^2 \quad \text{or}$$

$$\frac{R_{INPUT}(x = x_0)}{R_{INDUT}(x = 0)} = \left(\cos \left(\frac{\pi x_0}{1} \right) \right)^2.$$

This ratio of the input impedances is zero when $x_0 = \frac{1}{2} \cos \left(\frac{\pi}{2} \right) = 0..$ Typically, $x_0 = 0.45l..$

The quality factor of a rectangular planar patch antenna represents the various loss mechanisms and is given by

$$\frac{1}{Q_{TOTAL}} = \frac{1}{Q_{RADIATION}} + \frac{1}{Q_{OHMIC}} + \frac{1}{Q_{DIELECTRIC}} + \frac{1}{Q_{SURFACE\ WAVR_{\dot{c}}\dot{c}}}.$$

For very thin substrates, the expressions for the various quality factors are

$$Q_{OHMIC} = h\sqrt{\pi\mu\sigma f}, Q_{DIELECTRIC} = \frac{1}{\tan(\delta)}, \text{ and}$$

$$Q_{RADIATION} = \frac{2\omega\epsilon_r I \left[\dfrac{\int \{E\}^2 dA}{\displaystyle\oint \{E\}^2 dl}\right]}{hG_{TOTAL}},$$

where h, I, σ, ϵ_r are respectively the substrate thickness, patch length, patch conductivity, and substrate relative dielectric constant. $\frac{G_{TOTAL}}{1}$ is the conductance per unit length. The integral in the denominator is a closed loop integral over the perimeter of the patch, and the integral in the numerator is over the area of the patch.

The fractional bandwidth for a rectangular planar patch antenna is

$$\frac{2(f_2 - f_1)}{f_1 + f_2} = \frac{1}{Q_{TOTAL}}.$$

In case the voltage standing wave ratio is less than or equal to 2, and the reflection coefficient is less than 0.3, the bandwidth is

$$BW = 3.771 \left(\frac{\epsilon_r - 1}{\epsilon_r^2}\right) \left(\frac{hW}{\lambda_0 l}\right).$$

The electric and magnetic fields for a rectangular planar spiral inductor are

$$f(\theta, \phi) = \frac{\sin\left(\frac{kW}{2} \sin(\theta) \sin(\phi)\right)}{\frac{kW}{2} \sin(\theta) \sin(\phi)} \cos\left(\frac{kl}{2} \sin(\theta) \cos(\phi)\right)$$

where ϕ, θ are the poloidal and azimuthal angles in the spherical coordinate system. The electric and magnetic fields, at specific azimuthal and poloidal angles, are

$$F_E(\theta) = \cos\left(\frac{kl}{2} \sin(\theta)\right)\phi = 0 \text{ and } F_H(\theta) = \frac{\frac{kW}{2} \sin(\theta)}{\frac{kW \sin(\theta)}{2}} \cos(\theta)\phi = \frac{\pi}{2}.$$

The expressions for directivity take into account the two radiating slots individually. For the single slot case, with $k_0 h \ll 1$ the directivity can be expressed as

$$D_1 = \frac{4\pi U_{MAX}}{P_{RAD}} = \frac{\frac{4\pi}{2\eta_0}\left(\frac{\{V_9\}W}{\lambda_9}\right)^2}{\frac{1}{2\eta_0}\left(\frac{\{V_0\}^2}{\pi}\right)\int\left(\frac{\sin\left(0.5 k_0 W \cos\left(\theta\right)\right)^2}{\cos\left(\theta\right)}\right)^2(\sin\left(\theta\right))^3 d\theta}.$$

Similarly, for a two-radiation slot case, the expressions for directivity are

$$D_2 = \pi\left(\frac{2\pi W}{\lambda_0}\right)^2 \frac{1}{\int\left[\frac{\sin\left(\frac{k_0 W}{2}\cos\left(\theta\right)\right)^2}{\theta}\right](\sin\left(\theta\right))^3\left(\cos\left(\frac{k_0 l_e \sin\left(\theta\right)\sin\left(\phi\right)}{2}\right)\right)^2 d\theta d\phi}$$

where $0 \le \theta$, $\phi \le \pi$, I_e is the equivalent length.

The planar rectangular patch antenna used in cell phones and similar portable wireless equipment is the loaded planar patch antenna, called planar inverted F antenna (PIFA) (Fig. 2.22a, b, c).

Apart from the basic planar rectangular patch antenna, a number of interesting design variations have been experimented with, e.g., notched rectangular planar patch antenna. Planar patch antennas can be also be shaped as a circular disk or triangular patch. However, these are not used in real-world devices, as they are not easy to manufacture (compared to a rectangular patch) and do not make optimum use of available floor area.

Inverted F antennas can also be constructed without any shorting pin/wall, as shown in Fig. 2.22d, e.

2.6.2 Printed Planar Loop Antenna

The printed loop antennas [51] are used in small radio devices, e.g., modern RFICs use differential circuitry to provide improved common mode rejection; the inherent differential structure of a loop antenna interfaces well to such circuitry. Loop antennas are H-field radiators (unlike E-field radiators, e.g., monopole antennas). Loop antennas are immune to detuning because of hand or body effect. Loop antennas can be easily designed using printed circuit manufacturing methods, thus reducing costs. These antennas are small and rugged.

Unfortunately, they are difficult to design and implement, and so oftentimes existing designs are scaled appropriately to satisfy given set of specifications. This scheme is also prone to its own problems, and so extensive simulation is required to finalize a design.

The design and implementation scheme for a loop antenna is straightforward.

- Calculate required antenna dimensions using basic design equations, with knowledge of desired link range.

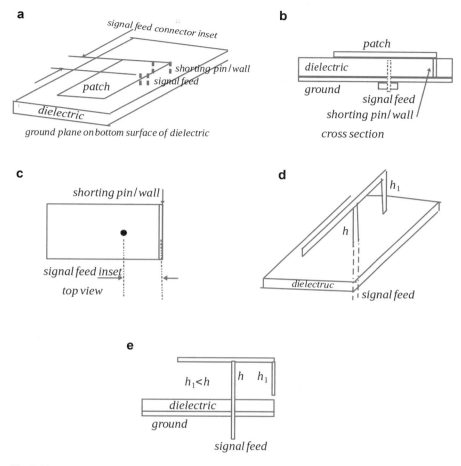

Fig. 2.22 (**a**) Planar inverted F antennas. (**b**) Planar inverted patch antenna cross section. (**c**) Planar inverted F antenna top view. (**d**) Inverted F antenna with no shorting pin—three-dimensional view. (**e**) Inverted F antenna with no shorting pin

- Calculation of tuning circuit components to force the antenna into resonance at the desired operating frequency.
- Simulation of proposed antenna geometry.
- Fabrication of printed circuit board with printed antenna structure.
- Laboratory measurement|testing of antenna resonant frequency and input impedance, including adjustment of discrete tuning capacitors to optimize resonant frequency and input impedance.

Unfortunately, simulation alone does not generate an accurate design as all CAD tools introduce simplifying approximations. The only way to design and implement an antenna is to prototype the device and incrementally improve the design.

The key characteristics of planar loop antennas are as follows:

- A loop antenna is a balanced structure and interfaces well to a differential input/ output circuit, such as a differential power amplifier output or a differential LNA input. The design should maintain physical symmetry of the loop structure.
- The input impedance of a loop antenna at natural resonance is high, ranging from 10 kΩ to 50 kΩ, a consequence of the loop antenna operating in a parallel-resonant mode at the operating frequency. This impedance value is much higher than the typical impedance of, e.g., power amplifier output or LNA input. Appropriate impedance matching circuits can transform the loop antenna impedance to a lower value. However, exact complex conjugate match is often not possible.
- The natural resonance of a loop antenna is narrowband, often only 5–10 MHz in bandwidth. The tuning of a loop antenna is influenced by nearby objects (i.e., hand effect or body effect). Some automated tuning circuitry is needed to tune the antenna back to resonance. A high-Q antenna provides attenuation of harmonic signal components, allowing the filtering required from discrete circuitry to be relaxed (or even eliminated).
- The natural resonant frequency of a loop antenna is inversely proportional to the size of the loop antenna, However, it is possible to use a discrete tuning components to tune a small loop antenna to resonance at frequencies below its natural resonant frequency. The discrete tuning components also provide a means by which the high native impedance is transformed to a lower and more useful value.
- The radiation efficiency of a loop antenna increases with size. If the designer has a choice in selecting the loop antenna size (assuming both may be tuned to resonance through the use of discrete tuning components), the larger antenna will generally provide better performance.

The design equations of a loop antenna are simple. The input parameters are the length of the sides and trace width and thickness. The effective radius corresponding to a rectangular cross-sectional trace of thickness t and width w is $b = 0.35t + 0.24w$. For a square loop, the loop inductance is

$$L = \frac{2\mu_0 l}{\pi}\left(\ln\left(\frac{l}{b}\right) - 0.774 \right) \text{ where } l = \sqrt{l_1}.$$

For a circular loop, the corresponding loop inductance is

$$L = \mu_0 l\left(\ln\left(\frac{8l}{b}\right) - 2.0 \right).$$

The radiation resistance of a small loop antenna is

$$R_{RAD} = \frac{320\pi^4 A^2 f^2}{v_{PHASE}^4}.$$

The radiation resistance is the only useful resistance in the loop antenna. The rest of the resistances just dissipate energy. The Ohmic trace resistance is

$$R_{TRACE} = \frac{1}{2w}\sqrt{\frac{\pi f \mu_0}{\sigma}}$$

where w, σ are the trace width and trace material conductivity, respectively. The capacitance of the resonant circuit is obtained from:

$$C = \frac{1}{4\pi^2 f^2 L}.$$

The antenna efficiency is given by

$$\eta = \frac{R_{RAD}}{R_{TRACE} + R_{SUBSTRATE} + R_{RAD} + R_{ESR}}.$$

The phase velocity of electromagnetic waves in the substrate dielectric is

$$V_{PGASE} = \frac{C}{\sqrt{\epsilon_{r,EFF}}}.$$

The equivalent electrical circuit for the planar loop antenna is shown in Fig. 2.23a, b.

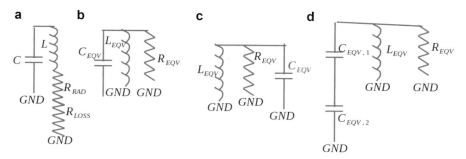

Fig. 2.23 (**a, b**) Loop antenna equivalent series and parallel lumped circuit element circuit. (**c, d**) Capacitively tapped impedance transformer corresponding to parallel lumped element circuit model for loop antenna

2.6.3   Planar Printed Dipole Antenna

The dipole antenna is the oldest antenna in use. Initially, dipole antennas were constructed with metal rods and pipes, of the appropriate length. The total length (sum of the two arms) is typically set to $\frac{\lambda}{2}$ though other lengths are also used, where λ is the operating frequency of the antenna. The dipole antenna exists in various forms, folded, array, etc., but the variant of interest here is the printed dipole, fabricated with printed circuit board technology.

A dipole antenna always needs a balancedldifferential input [50] for proper operation, and the balun (balanced-unbalanced) is a $\frac{\lambda}{4}$ transmission line transformer. For a printed dipole antenna, an integrated balun consist of a microstrip line, $\frac{\lambda_g}{4}$ open stub (λ_g is the guided wavelength of the microstrip line), and a $\frac{\lambda_s}{4}$ short-circuited slot line (λ_S guided wavelength of the slot line). The strip conductor and the ground plane of the microstrip line in combination are equivalent to the inner conductor and outer conductor of a coaxial line, respectively. The slot line is equivalent to any two-conductor transmission line. The $\frac{\lambda_g}{4}$ open circuit transmission line stub acts as a shorting circuit at the feed point. The printed dipole with the integrated balun exhibits broadband behavior (40%) and is used in wireless communications and antenna arrays. However, the printed dipole with an integrated balun whose feed point is fixed at its top (fixed integrated balun) has to be designed at the dipole's resonant resistance (80Ω), so a 63 Ω quarter wavelength transformer is required to match with 50 Ω testing equipment. The bandwidth of the printed dipole with fixed integrated balun degrades (30%–50%) when directly connected to a 50 Ω input, because of impedance mismatch.

Therefore, to curb the use of a quarter wavelength transformer and the bandwidth reduction, a simple modification is needed. Although a printed Yagi-Uda dipole or a T-dipole can optimize these problems, their manufacture is problematic—the printed Yagi-Uda dipole with integrated balun needs a high dielectric constant substrate for surface wave enhancement, while the T-dipole has a large size.

To overcome these issues, any printed dipole with integrated balun can be matched to a 50-feed by adjusting the feed point of the integrated balun. As a result short-circuited slot line gets divided into a slot line and a shorted stub, thereby forming a single stub tuning circuit with a shunt shorted stub. This immediately converts a high impedance to a low impedance. Since the position of the feed point is adjustable, the adjusted integrated balun can match different impedance values, which is extremely useful for antenna arrays. The impedance matching scheme is basically the familiar load-source pull technique used widely to experimentally optimize impedances. Figure 2.24a, b shows the original and modified impedance matching scheme.

With reference to Fig. 2.24c, the printed dipole is fed by a slot line which is coupled to a microstrip line, and extended to a shorted stub. The center-fed input impedance of the printed dipole without the integrated balun is Z_d. The impedance looking into the slot line at the feed point is Z_C given by

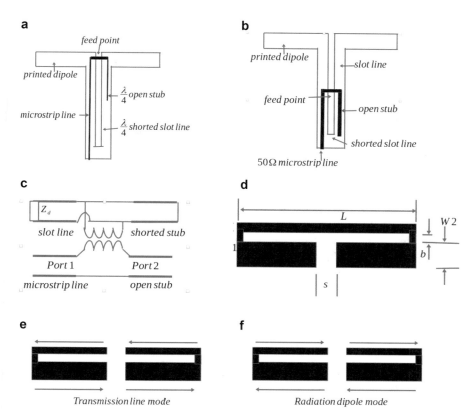

Fig. 2.24 (**a**) Planar microstrip-based balun impedance matched dipole antenna. (**b**) Modified planar microstrip-based balun impedance matched dipole antenna. (**c**) Mixed lumped element and transmission line circuit model corresponding to Fig. 2.24b. (**d**) Asymmetric folded dipole antenna. (**e**) Asymmetric folded dipole antenna transmission line mode of operation. (**f**) Asymmetric folded dipole antenna radiation dipole mode of operation

$$Z_C = \frac{Z_{0,SLOT}\left(Z_D + jZ_{0,SLOT}\tan\left(\beta_{SLOT}l_{SLOT}\right)\right)}{Z_{0,SLOT} + jZ_D\tan\left(\beta_{SLOT}l_{SLOT}\right)}$$

where the terms are self-explanatory.

The impedance of Z_C in parallel with the shorted stub is Z_B

$$Z_B = \frac{jZ_C Z_{0,SLOT}\tan\left(\beta_{SLOT}l_{SHORT\ STUB}\right)}{Z_C + jZ_{0,SLOT}\tan\left(\beta_{SLOT}l_{SHORT\ STUB}\right)}.$$

The slot line and microstrip line coupling is represented as an ideal transformer with a turns ratio of n:1, such that the impedances are transformed as $Z_A = n^2 Z_B$..

The input impedance looking into the microstrip line and at the signal feed point is

$$Z_{INPUT} = Z_A - jZ_{0,OPEN\ STUB} \cot\left(\beta_{OPEN\ STUB} l_{OPEN\ STUB}\right).$$

As the dipole antenna has been in use for a long time, the planar printed dipole antenna's physical shape has been modified to satisfy space constraints. One popular version is the folded dipole antenna [52] (one variant of which is the strip asymmetric folded antenna) (Fig. 2.24d). The strip asymmetric folded dipole's geometricl physical dimensions L, b, d, s, W1, and W2 must be adjusted experimentally to achieve correct input impedance matching and target bandwidth. The antenna has no ground plane, enabling a radiation pattern that is similar to that of a simple dipole of the same length l. On the other hand, the input impedance is four times larger compared to that of a conventional dipole $1 \leq \frac{\lambda}{2}$. The length of a single-wire dipole is typically $\frac{\lambda}{4} \leq 1 \leq \frac{\lambda}{2}$ for directivity side lobe-free directivity, and s < 0.002λ. The separation distance b between the two strip transmission lines of the folded dipole should not exceed 0.05λ..

The analysis of this antenna uses the transmission line mode, and the unbalanced radiation antenna mode (Fig. 2.24e). The input impedance of this antenna is

$$Z_{INPUT} = \frac{2(1+\gamma)^2 Z_D Z_T}{(1+\gamma)^2 Z_D + 2Z_T}$$

where Z_D, Z_T are the input impedances of the dipole and transmission line

$$Z_T = \frac{120\pi}{\sqrt{\epsilon_r}} \frac{K(k)}{K^{PRIME}(k)} \tan\left(\frac{\beta L}{2}\right)$$

where K, K^{PRIME} are elliptic functions

$$\gamma = \frac{\ln\left(4C + 2\sqrt{4C^2 - \left(\frac{W_1^2}{2}\right)}\right) - \ln\left(W_1\right)}{\ln\left(4C + 2\sqrt{4C^2 - \left(\frac{W_2^2}{2}\right)}\right) - \ln\left(W_2\right)}$$

where 2C is the distance between the middle of one dipole strip and middle of the second, and is expressed as

$$C = \frac{b}{2} + \frac{W_1}{4} + \frac{W_2}{4}.$$

Clearly analytically calculating and using these expressions for the impedances of the microstrip transmission line and antenna are difficult, so a simplified model of the same device is required. This is achieved by first computing the input impedance of the asymmetric folded antenna in the transmission line, followed by calculating the input impedance of the same antenna in the radiation mode, and combining the results. In the transmission line mode, the impedance is

$$Z_T = Z_0 \left(\frac{Z_L + jZ_0 \tan \left(\frac{\beta l}{2} \right)}{Z_0 + jZ_L \tan \left(\frac{\beta l}{2} \right)} \right)$$

where β, I, Z_0, Z_L are respectively the propagation constant, length, characteristic impedance, and load impedance.

To compute the input impedance in the radiation mode, a key simplifying assumption is that the asymmetric folded dipole can be regarded as a symmetric two-port network, and then using the concepts of evenlodd mode transmission line propagation with the electric and magnetic walls. In the even mode, the admittances of the two halves of the antenna (separated at the magnetic wall) are

$$Y_{1,2} \left(\frac{L}{2} \right) = Y_{d1.2} \left(\frac{Y_L + jY_{d1,2} \tan \left(\frac{\beta L}{2} \right)}{Y_{d1.2} + jY_L \tan \left(\frac{\beta L}{2} \right)} \right),$$

and for the odd radiation mode, the admittances are

$$Y_{O1.2} \left(\frac{L}{2} \right) = jY_{d1.2} \tan \left(\frac{\beta L}{2} \right).$$

In the even mode, with L tending to infinity,

$$Y_{e1.2} \left(\frac{L}{2} \right) = -jY_{d1.2} \cot \left(\frac{\beta L}{2} \right) \quad \text{and so}$$

$$Y_e = Y_{e1} + Y_{e2} \text{ and } Y_0 = Y_{01} + Y_{02}$$

Then

$$Z_{DIPOLE} = \frac{1}{Y_o - Y_e}.$$

Another popular variant of the basic microstrip dipole antenna is the wide band fabricated using common printed circuit board manufacturing technology. The key design trick underlying the broadwide band planar microstrip dipole antenna is it increases the width of the dipole strips. Some common variants of microstrip wide band dipole antenna are shown in Fig. 2.25a, b.

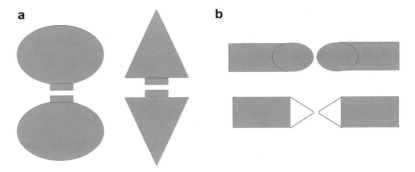

Fig. 2.25 (**a**) Microstrip dipole wide band antenna ellipse and diamond shaped. (**b**) Microstrip wide band dipole antenna

2.6.4 Printed Planar Antenna Surface Wave Issues and Remedies

As printed planar antennas are very easy to manufacture and integrate with RF| microwave circuits, they are used in novel ways. These transmission line-based antennas can integrate three-terminal devices and are mechanically rigid and the metallic ground planes serve as heat sinks. Most importantly, signals can be fed into or retrieved from these antennas using similar transmission lines. The method used to feed signals into or retrieve signals from these antennas is important, as number of key performance characteristics as antenna cross-polarization, patterns, and bandwidth.

Unfortunately, the grounded dielectric slabs on which microstrip and coplanar waveguide (CPW) are fabricated support TM0 surface waves [54], leading to energy loss, thus decreasing efficiency. The losses are small at lower frequencies, but significant at microwave and millimeter-wave frequencies. The thickness of the substrate and permittivity and frequency of operation determine the amount of surface wave losses. The radiation characteristics of printed planar antennas vary. Patch and resonant slot antennas demonstrate broad, low gain patterns that make them suitable for use in multielement beamforming arrays or portable handsets. Some of these antennas can be modified easily for dual-linear or circular polarization. More sophisticated classes demonstrate higher gains and some are capable of frequency scanning.

One way to reduce surface waves is to make the antenna substrate electrically thick, but this reduces Q value of the antenna cavity for increased bandwidth. Unfortunately, unless an optimum thickness is used, high levels of TM0 surface waves can reduce the radiation frequencies and the radiation pattern if the surface wave generates radiation of its own (at the edge of finite ground antennas). The thick substrate surface wave issue is also there for high-frequency antennas on high permittivity substrates.

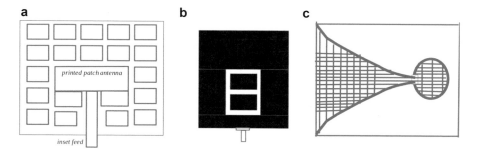

Fig. 2.26 (**a**) Photonic bandgap scheme to curb printed planar patch antenna surface waves. (**b**) Planar folded slot antenna. (**c**) Tapered Vivaldi planar slot antenna. Hatched regions indicate bare dielectric surface

Another method to curtail surface waves involves micromachining closely spaced holes into the high dielectric constant substrate to lower the effective permittivity in the antenna region and, therefore, reduce the electrical thickness. Although this method will increase radiation efficiency, some bandwidth degradation will occur, as a result of lowered effective permittivity.

The latest suggested scheme to control surface waves is to exploit photonic bandgap (PBG). In this case, a periodic array of perturbations suppresses the surface wave mode (Fig. 2.26a), which shows the layout of this scheme as applied to an inset fed patch antenna. It consists of capacitive pads on the top plane connected to ground with a shorting pin, one per. This periodic structure eliminates TM0 surface waves, and vastly boosts antenna gain, compared to a reference antenna lacking PBG. In addition, PBG-enhanced antennas also have enhanced bandwidth.

2.6.5   Novel Printed Planar Antennas

A slot antenna [54] is a narrow slit in a ground plane. It is a very versatile device, such that with some modifications input signals can be fed via waveguide, CPW, coaxial cable, slot line, or microstrip transmission lines and used widely in wireless communication and radar systems.

The resonant half-wavelength slot antenna is preferred owing to its compact size, but suffers from large input impedance, about $300\,\Omega$ or higher which makes impedance matching difficult. This issue can be circumvented by using an offset microstrip feed or the folded slot antenna, which is related to the folded dipole (examined earlier) by Booker's relation. By folding the slot on itself, such that its overall length remains approximately a half-wavelength, increasing the number of folds reduces the radiation resistance.

The CPW version of the folded slot antenna requires no input matching, which makes it very easy to integrate with microwave circuits. The broad radiation pattern also makes this antenna very attractive for wireless communication systems, with carrier frequencies in the 10s of GHz.

The basic structure for the microstrip fed folded slot antenna is shown in Fig. 2.26b. The folded slot antenna is etched in the ground plane of the substrate. One of the inner metallizations is connected to the microstrip in the top plane with a shorting pin. An input impedance of 50 Ω is easily obtained for a twofold slot with appropriate choice of parameters as dielectric permittivity, thickness, etc. Using finite domain time domain (FDTD) simulations, it is seen that the feeding structure operates effectively and the perpendicular fields exist at the edges of the slot, contributing to the cross-polarization of the radiated fields. Also, reducing the area at the edges of the slot reduces both the cross-polarization and bandwidth.

The tapered slot antenna is a planar device, with an endfire radiation pattern, and achieves high directivity and/or bandwidth. It can be used for millimeter-wave imaging, power combining, and integrated antenna element. This antenna is etched into the metallization on one side of a low dielectric constant substrate. Radiation at a particular frequency occurs when the slot is a certain diameter, so that the starting slot width should be at least one-half wavelength for efficient radiation—maximum and minimum widths determine its bandwidth. It is fabricated on thin, low dielectric constant substrates to achieve good radiation patterns and maintain targeted radiation efficiency, both of which are reduced by TM0 surface wave losses. As with the patch antenna, micromachining reduces the electrical thickness of millimeter-wave by reducing the effective permittivity, and allows a physically thicker substrate, which provides needed mechanical rigidity. A tapered slot antenna is shown in Fig. 2.26c.

Resonance based (standing wave) antennas, e.g., patch and slot antennas, have narrow bandwidth and modest gain radiation patterns. Another class of antennas— traveling wave or **nonresonant** antennas—radiates as the wave propagates through the structure. These structures are electrically large compared to resonance-based structures and demonstrate higher gain. When fabricated correctly, they also produce wide bandwidth radiation. Traveling wave antennas can be of two types—leaky wave and surface wave. **The leaky wave antenna radiates continuously as the wave propagates. The surface wave antenna radiates at discontinuities in the structure which disrupt the propagating surface wave and is a slow wave structure**.

This type of leaky wave antenna utilizes a higher-order radiative microstrip mode to operate. The radiation mode has odd symmetry and peaks at the edges of the microstrip cross section. The zero of this mode occurs at the center. Therefore, as the signal propagates along the microstrip transmission line, energy radiates continuously along the edges, creating the main radiation beam. In case the end of the transmission is open circuited, the remainder of the energy is reflected back generating a second beam. Therefore, in non-terminated long antennas, the reflected power is small, resulting in a weaker second beam. More importantly, the direction of the main beam is a function of frequency, allowing frequency scanning. The microstrip leaky wave radiates efficiently over a fixed bandwidth, limiting maximum scanning. Moreover, feeding and length of the antenna also have some effect on the pattern. Signal injection into this antenna is tricky and various schemes have been tried out. The higher-order leaky wave mode must be excited with minimal excitation of the fundamental (even) mode. Often periodic structure of transverse slots is used to suppress excitation of the fundamental even mode.

2.7 Conclusion

A detailed examination of planar spiral inductors, embedded|integrated (on-chip) planar transformers consisting of planar spiral inductors (co-planar or vertically stacked), and common planar antennas has been presented. The next chapter contains analysis and design examples of these electronic circuits.

References

1. Long, J. R., & Cop. (1997). In addition, a periodic structure of transverse slots is used to suppress excitation of the fundamental even mode. Figure 9 shows the cross section of the antenna's electric field profile with and without the mode suppressor. In the top figure, a substantial amount of asymmetry in the vertical field can be observed, indicating that both the fundamental (even) and the leakyeland, M. A. The Modeling, Characterization and Design of Monolithic Inductors for Silicon RF ICs. *IEEE Journal of Solid-State Circuits, 32*, 357–369.
2. Niknejad, A. M., & Meyer, R. G. (1998). Analysis, design, and optimization of spiral inductors and transformers for Si RF ICs. *IEEE Journal of Solid-State Circuits, 33*, 1470–1481.
3. Reyes, A. C., El-Ghazaly, S. M., Dorn, S. J., Dydyk, M., Schrider, D. K., & Patterson, H. (1995). Coplanar waveguides and microwave inductors on silicon substrates. *IEEE Transactions on Microwave Theory and Technology, 43*, 2016–2022.
4. Ashby, K. B., Koullias, I. C., Finley, W. C., Bastek, J. J., & Moinian, S. (1996). High Q inductors for wireless applications in a complementary silicon bipolar process. *IEEE Journal of Solid-State Circuits, 31*, 4–9.
5. Lu, L. H., Ponchak, G. E., Bhattacharya, P., & Katehi, L. (2000). High-Q X-band and K'-band micromachined spiral inductors for use in Si-based integrated circuits. *Proceedings of Silicon Monolithic Integrated Circuits RF Systems*, 108–112.
6. Bahl, I. J. (1999). Improved quality factor spiral inductor on GaAs substrates. *IEEE Microwave Guided Wave Letters, 9*, 398–400.
7. Ribas, R. P., Lescot, J., Leclercq, J. L., Bernnouri, N., Karam, J. M., & Courtois, B. (1998). Micromachined planar spiral inductor in standard GaAs HEMT MMIC technology. *IEEE Electron Device Letters, 19*, 285–287.
8. Takenaka, H., & Ueda, D. (1996). 0.15µm T-shaped gate fabrication for GaAs MODFET using phase shift lithography. *IEEE Transactions on Electron Devices, 43*, 238–244.
9. Chiou, M. H., & Hsu, K. Y. J. (2006). A new wideband modeling technique for spiral inductors. *IET Microwave, Antennas, and Propagation, 151*, 115–120.
10. Lu, H.-C., Chan, T. B., Chen, C. C. P., & Liu, C. M. (2010). Spiral inductor synthesis and optimization with measurement. *IEEE Transactions on Advanced Packaging, 33*.
11. Talwalkar, N. A., Yue, C. P., & Wong, S. S. (2005). Analysis and synthesis of on-chip spiral inductors. *IEEE Transactions on Electron Devices, 52*, 176–182.

12. Mukherjee, S., Mutnury, S., Dalmia, S., & Swaminathan, M. (2005). Layout-level synthesis of RF inductors and filters in LCP substrate for Wi-fi applications. *IEEE Transactions on Microwave Theory and Technology, 53*, 2196–2210.
13. Kulkarni, J. P., Augustine, C., Jung, C., & Roy, K. (2010). Nano spiral inductors for low-power digital spintronic circuits. *IEEE Trans. on Magnetics, 46*, 1898–1901.
14. Greenhouse, H. M. (1974). Design of planar rectangular microelectronic inductors. *IEEE Transactions on Parts, Hybrids and Packaging, 10*, 101–109.
15. https://books.google.co.in/books?hl=en&lr=&id=K3KHi9lIltsC&oi=fnd&pg=PR13& dq=grover+inductance+calculations&ots=dPYlK2rxOd&sig=ZLMajbfyFc0P4 EBh5wukPJZBa8w#v=onepage&q=grover%20inductance%20calculations&f=false
16. Jenei, S., Nauwelaers, B. K. J. C., & Decoutere, S. (2002). Physics-based closed-form inductance expression for compact modeling of integrated spiral inductors. *IEEE Journal of Solid-State Circuits, 37*, 77–80.
17. Asgaran, S. (2002). New accurate physics-based closed-form expressions for compact modeling and design of on-chip spiral inductors. *Proceedings of the 14th International Conference on Microelectronics*, 247–250.
18. Mohan, S. S., Hershenson, M. M., Boyd, S. P., & Lee, T. H. (1999). Simple accurate expressions for planar spiral inductance. *IEEE Journal of Solid-State Circuits, 34*, 1419–1424.
19. Chen, C. C., Huang, J. K., & Cheng, Y. T. (2005). A closed-form integral model of spiral inductor using the Kramers-Kronig relations. *IEEE Microwave and Wireless Components Letters, 15*.
20. Sieiro, J., Lopez-Villegas, J. M., Cabanillas, J., Osorio, J. A., & Samitier, J. (2002). A physical frequency-dependent compact model for RF integrated inductors. *IEEE Transactions on Microwave Theory and Technology, 50*, 384–392.
21. Sun, H., Liu, Z., Zhao, J., Wang, L., & Zhu, J. (2008). The enhancement of Q-factor of planar spiral inductor with low-temperature annealing. *IEEE Transactions on Electron Devices, 55*, 931–936.
22. Tsai, H. S., Lin, L., Frye, R. C., Tai, K. L., Lau, M. Y., Kossives, D., Hrycenko, F., & Chen, Y. K. (1997). Investigation of current crowding effect on spiral inductors. *IEEE MTT-S Symposium on Technologies to Wireless Applications*, 139–142.
23. Bushyager, N., Davis, M., Dalton, E., Laskar, J., & Tentzeris, M. (2002). Q-factor and optimization of multilayer inductors for RF packaging microsystems using time domain techniques. *Electronic Components and Technology Conference*, 1718–1721.
24. Eroglu, A., & Lee, J. K. (2008). The complete design of microstrip directional couplers using the synthesis technique. *IEEE Transactions on Instrumentation and Measurement, 12*, 2756–2761.
25. Costa, E. M. M. (2009). Parasitic capacitances on planar coil. *Journal of Electromagnetic Waves and Applications, 23*(17–18), 2339–2350.
26. Nguyen, N. M., & Meyer, R. G. (1990). Si IC-compatible inductors and LC passive filter. *IEEE Journal of Solid-State Circuits, 27*(10), 1028–1031.
27. Zu, L., Lu, Y., Frye, R. C., Law, Y., Chen, S., Kossiva, D., Lin, J., & Tai, K. L. (1996). High Q-factor inductors integrated on MCM Si substrates. *IEEE Transactions on Components. Packaging and Manufacturing Technology, Part B: Advanced Packaging, 19*(3), 635–643.
28. Burghartz, J. N., Soyuer, M., & Jenkins, K. (1996). Microwave inductors and capacitors in standard multilevel interconnect silicon technology. *IEEE Transactions on Microwave Theory and Technology, 44*(1), 100–103.
29. Merrill, R. B., Lee, T. W., You, H., Rasmussen, R., & Moberly, L. A. (1995). Optimization of high Q integrated inductors for multi-level metal CMOS. *IEDM*, 38.7.1–38.7.3.
30. Chang, J. Y. C., & Abidi, A. A. (1993). Large suspended inductors on silicon and their use in a 2 μm CMOS RF amplifier. *IEEE Electron Device Letters, 14*(5), 246–248.
31. Craninckx, J., & Steyaert, M. (1997). A 1.8-GHz low-phase-noise CMOS VCO using optimized hollow spiral inductors. *IEEE Journal of Solid-State Circuits, 32*(5), 736–745.

32. Lovelace, D., & Camilleri, N. (1994). Silicon MMIC inductor modeling for high volume, low cost applications. *Microwave Journal*, 60–71.
33. Kamon, M., Tsulk, M. J., & White, J. K. (1994). FASTHENRY a multipole accelerated 3-D inductance extraction program. *IEEE Transactions on Microwave Theory and Technology, 42*(9), 1750–1757.
34. Pettenpaul, E., Kapusta, H., .Weisgerber, A., Mampe, H., Luginsland, J., Wolff, I. (1988). CAD models of lumped elements on GaAs up to 18 GHz, *IEEE Transactions of Microwave Theory and Technology, 36*(2) 294–304.
35. Howard, G. E., Yang, J. J., & Chow, Y. L. (1992). A multipipe model of general strip transmission lines for rapid convergence of integral equation singularities. *IEEE Transactions on Microwave Theory Technology, 40*(4), 628–636.
36. Gharpurey, R. *Modeling and Analysis of Substrate Coupling in Integrated Circuits Doctoral.* Thesis, University of California.
37. Stetzler, T., Post, I., Havens, J., & Koyama, M. (1995). A 2.7V to 4.5V single-chip GSM transceiver RF integrated circuit. *IEEE International Solid-State Circuits Conference*, 150–151.
38. Kim, B. K., Ko, B. K., Lee, K., Jeong, J. W., Lee, K.-S., & Kim, S. C. (1995). Monolithic planar RF inductor and waveguide structures on silicon with performance comparable to those in GaAs MMIC. *IEDM*, 29.4.1–29.4.4.
39. Krafesik, D., & Dawson, D. (1986). A closed-form expression for representing the distributed nature of the spiral inductor. *Proceedings of the IEEE-MTT Monolithic Circuits Symposium*, 87–91.
40. Kuhn, W. B., Elshabini-Riad, A., & Stephenson, F. W. (1995). Centre-tapped spiral inductors for monolithic bandpass filters. *Electronics Letters, 31*(8), 625–626.
41. Derkaoui, M. (2019). Modeling and simulation of an integrated planar transformer for RF systems. *International Journal of Industrial and Manufacturing Systems and Engineering, 4*(6), 54–63. ISSN: 2575-3150 (Print); ISSN: 2575-3142 (Onlinehttps://www.amazon.in/Microwave-Engineering-4ed-David-Pozar/dp/8126541903)
42. Pozar, D. M. *Microwave engineering.* 4th ed. https://www.amazon.in/Microwave-Engineering-4ed-David-Pozar/dp/8126541903
43. Mohri, K., Uchitama, T., Panina, L. V., Yamamoto, M., & Bushida, K. Recent Advances of Amorphous Wire CMOS IC Magneto-Impedance Sensors: Innovative High-Performance Micromagnetic Sensor Chip Copyright © 2015 Kaneo Mohri et al. This is an open access article distributed under the Creative Commons Attribution License, which permits unrestricted use, distribution and reproduction in any medium, provided that it is properly cited.
44. Kurup, H. B., Dinesh, S., Ramesh, M., & Rodrigues, M. (2020). Low profile dual-frequency shorted patch antenna. *International Journal of recent Technology and Engineering, 8*(5), 2277–3878.
45. Mishra, A., Singh, P., Yadav, N. P., Ansari, J. A., & Viswakarma, B. R. (2009). Compact shorted microstrip patch antenna for dual band operation. *Progress In Electromagnetics Research C, 9*, 171–182.
46. Ansari, J. A., Singh, P., Yadav, N. P., & Viswakarma, B. R. (2009). Analysis of shorting pin loaded half disk patch antenna for wideband operation. *Progress in Electromagnetics Research C, 6*, 179–192.
47. Kathiravan, K., & Bhattacharyya, A. K. (1989). Analysis of triangular patch antennas. *Electromagnetics, 9*(4), 427–438. https://doi.org/10.1080/02726348908915248
48. Malik, J., & Kartikeyan, M. (2011). A stacked equilateral triangular patch antenna with Sierpinski gasket fractal for WLAN applications. *Progress In Electromagnetics Research Letters, 22*, 71–81. https://doi.org/10.2528/PIERL10122304. http://www.jpier.org/PIERL/pier.php?paper=10122304
49. Tripathi, A. K., Bhatt, P. K., & Pandey, A. K. (2012). A comparative study of rectangular and triangular patch antenna using HFSS and CADFEKO. *International Journal of Computer Science and Information Technologies, 3*(6), 5356–5358.

50. Li, R. L., Wu, T., Pan, B., Lim, K., Laskar, J., & Tentzeris, M. M. (2009). Equivalent circuit analysis of a broadband printed dipole with adjusted integrated balun and array for base station applications. *IEEE Transactions on Antennas and Propagation, 57*(7).
51. Application Note 639 Design of Printed Trace Differential Loop Antennas Copyright 2021 Silicon Laboratories Inc.
52. Hua, G., Yang, C., Lu, P., Zhou, H., & Peng, W. Microstrip folded dipole antenna. *International Journal of Antennas and Propagation, 2013*, 603654, 6 https://doi.org/10.1155/2013/603654 Hindawi Publishing Corporation.
53. Balanis, C. A. (2016). *Antenna theory analysis and design fourth edition*. Wiley. Library of Congress Cataloging-in-Publication Data:ISBN 978-1-118-642060-1 (cloth) 1. Antennas (Electronics) I. Title.TK7871.6.B354 2016 621.382.
54. https://www.microwavejournal.com/articles/2677-planar-integrated-antenna-technology

Chapter 3
SPICE Based Design and Analysis of Planar Spiral Inductors and Embedded|Integrated Planar Spiral Inductor Transformers and Planar Antennas

3.1 Single Planar Microstrip Inductor on Printed Circuit Board

The supplied C computer language [1] executable *planarind* accepts the following inductor parameters as input:

- Characteristic impedance of the inductor
- Dielectric constant of the printed circuit board substrate material
- Mid-band frequency of the inductor's operating frequency range
- Target inductance value
- Substrate thickness
- Trace conductor thickness
- Trace width
- Mean radius (nonzero only for circular loop inductor)
- Tolerance—how close does the calculated inductance value need to be with reference to the supplied target value

The supplied C computer language [1] executable planarind uses these inputs to compute the values of components of the equivalent electrical circuit of the planar inductor, and arranges these values as a SPICE [ ] text input format netlist that can be used with any available open-source or proprietary SPICE[] simulator. Typing **./planar**ind at the Linux|MinGW shell prompt generates the following help information:

Please note that in all subsequent sections of this chapter, all computer-generated text are in bold font.

Supplementary Information The online version contains supplementary material available at [https://doi.org/10.1007/978-3-031-08778-3_3].

© The Author(s), under exclusive license to Springer Nature Switzerland AG 2023
A. Banerjee, *Planar Spiral Inductors, Planar Antennas and Embedded Planar Transformers*, https://doi.org/10.1007/978-3-031-08778-3_3

```
incorrect|insufficient arguments
batch|command line argument mode
single microstrip straight segment and single microstrip loop
./planarind b|B|c|C 1|2
<characteristic impedance(Ohm)>
<substrate dielectric constant>
<mid-band frequency(MHz)>
<target inductance(nH)>
<substrate thickness(mm)>
<trace thickness(mm)>
<trace width(mm)>
<tolerance on calculation>
*****************
1 - single segment microstrip
2 - single microstrip loop
sample command line input for single microstrip inductor
./planarind b 1 50 4.75 500 5 1.5 0.01 0.75 0 0 0.95
sample command line input for single microstrip loop inductor
./planarind C 2 50 4.75 500 7.6 1.5 0.01 0.75 0 0 0.95
```

There are two types of SPICE [2–5] analysis that are commonly used—steady-state or transient analysis and start-up or AC (small signal) analysis. The SPICE [] text netlist generated by *planarind* using the supplied sample batch|command line argument input for a single rectangular microstrip inductor on a printed circuit board is:

```
./planarind b 1 50 4.75 500 5 1.5 0.01 0.75 0 0 0.95
microstrip length 0.050000 m
calculated inductor 1.390185e-08 H
SPICE netliat planarind.cir
```

The generated SPICE [2–5] netlist is:

```
SINGLE MICROSTRIP RECTANGULAR INDUCTOR

.PARAMS C=1.231771e-12 FREQ=5.000000e+08 L=1.390185e-08 R=0.1120
.PARAMS FLLIM=5.000000e+04 FHLIM=1.0E+9 AMPL=5 CT=1.0E-15 RR=50.0
RSMALL=0.001 TS='1.0/(15.0*FREQ)' TSTOP='20.0/FREQ' TSTRT='1.0/
FREQ' TOT=100000

** ADJUST TEST CAPACITOR VALUE
** FOR BETTER RESONANCE

.SUBCKT PLINDMSS 1 2
** 1 IN
** 2 OUT
C0 1 0 {C}
L0 1 3 {L}
R0 3 2 {R}
.ENDS
```

```
** TRANSIENT ANALYSIS
R0 1 2 {RSMALL}
VTST0 2 3 DC 0.0 AC 0.0
*XPLINDMSS 3 0 PLINDMSS
*VSIG 1 0 DC 0.001 SIN(0 {AMPL} {FREQ} 0 0 0)


** RESONANCE MEASUREMENT
CT 3 4 {CT}
RS 1 2 {RR}
RL 4 0 {RR}


XPLINDMSS 2 3 PLINDMSS
VSIG 1 0 DC 0.001 AC {AMPL}


** S PARAMETER
RS 1 2 {RR}
RL 3 0 {RR}


** S11 S21
XPLINDMSS 2 3 PLINDMSS
** S22 S12
XPLINDMSS 3 2 PLINDMSS
VSIG 1 0 DC 0.001 AC 1.0


.OPTIONS METHOD=GEAR NOPAGE RELTOL=1m
** TRANSIENT ANALYSIS
*.IC
*.TRAN {TS} {TSTOP} {TSTRT} UIC
*.PRINT TRAN V(3) I(VTST0)


** RESONANCE, S PARAMETER
.AC LIN {TOT} {FLLIM} {FHLIM}
** RESONANCE MEASUREMENT
.PRINT AC V(4)


** S PARAMETER S11 S22
*.PRINT AC V(2)
** S PARAMETER S21 S12
.PRINT AC V(3)


.END
```

By commenting out appropriate lines in the above SPICE [2–5] netlist, it can be used alternatively for transient analysis, AC (small signal) resonance estimation, and AC (small signal) S parameter measurement.

The key inductor performance metric is the quality (Q) factor, and this is estimated in this case by commenting out all lines in the above netlist related to AC (small signal) analysis (resonance and S parameter) and executing a new SPICE [2–5] netlist, each corresponding to a different frequency in the selected frequency band. *This is essential, as the series resistance in the inductor's equivalent electrical circuit is frequency dependent—even if that difference appears in the fifth or sixth decimal place.* The selected frequency range in this case is 50 MHz–2 GHz. The underlying steps of the Q factor estimation is simple: the impedance of the inductor is $Z = R(f) + j\omega L |Z| = \frac{V_{RMS}}{I_{RMS}}$ where V_{RMS}, I_{RMS} are respectively the RMS (root mean square) voltage across and current through the inductor. For each frequency, R (f) and L are calculated by *planarind*, so that

$$Q = \frac{\omega L}{R(f)} \quad L = \frac{Z - R(f)}{j\omega}$$

Using this simple scheme along with the SPICE [2–5] netlist (regenerated for each frequency in the selected frequency band), the values for the RMS (root mean square) voltage, current, and the frequency-dependent series resistance of the planar microstrip inductor are listed as follows, where the first value in each row is the frequency for which the corresponding voltage across and current through the planar microstrip inductor were made. *The last number in each row is the frequency-dependent series resistance—although this value appears the same for each row, it is because the skin depth effect is very small, and does not alter the series resistance noticeably.* **The Q factor measurement uses the steady-state analysis feature of SPICE [2–5]—transient analysis:**

```
5.0E+7,3.523181e+00,8.511896e-01,0.1120
1.0E+8,3.517104e+00,4.723874e-01,0.1120
1.5E+8,3.517109e+00,3.368641e-01,0.1120
2.0E+8,3.517111e+00,2.636655e-01,0.1120
2.5E+8,3.517112e+00,2.167356e-01,0.1120
3.0E+8,3.517113e+00,1.837066e-01,0.1120
3.5E+8,3.517113e+00,1.590391e-01,0.1120
4.0E+8,3.517303e+00,1.399021e-01,0.1120
4.5E+8,3.511464e+00,1.242868e-01,0.1120
5.0E+8,3.511342e+00,1.115666e-01,0.1120
5.5E+8,3.511348e+00,1.009418e-01,0.1120
6.0E+8,3.511374e+00,9.191924e-02,0.1120
6.5E+8,3.511393e+00,8.415658e-02,0.1120
7.0E+8,3.511397e+00,7.741807e-02,0.1120
7.5E+8,3.511795e+00,7.156092e-02,0.1120
8.0E+8,3.505798e+00,6.628498e-02,0.1120
8.5E+8,3.505702e+00,6.174006e-02,0.1120
9.0E+8,3.505736e+00,5.775373e-02,0.1120
9.5E+8,3.505760e+00,5.425624e-02,0.1120
1.0E+9,3.505780e+00,5.119971e-02,0.1120
1.1E+9,3.505808e+00,4.627402e-02,0.1120
1.2E+9,3.505813e+00,4.276175e-02,0.1120
1.3E+9,3.505817e+00,4.050277e-02,0.1120
1.4E+9,3.505770e+00,3.935407e-02,0.1120
1.5E+9,3.505772e+00,3.915165e-02,0.1120
1.6E+9,3.505774e+00,3.974022e-02,0.1120
1.7E+9,3.505775e+00,4.096286e-02,0.1120
1.8E+9,3.505777e+00,4.267814e-02,0.1120
```

The supplied C computer language [1] executable **qfactmeas** accepts the above data as input, and generates the Q factor value for each frequency, as shown in Fig. 3.1 using

$$Q = \frac{\omega L}{R(f)} = \frac{\dfrac{-j\omega^2 (Z - R(f))}{\omega^2}}{R(f)} = \frac{-j(Z - R(f))}{R(f)}$$

The RMS values for the current and voltage are generated using the simple C computer language[] executable **rmscalc**.

The AC (small signal) feature of SPICE [2–5] is used to measure a number of key performance features of the rectangular planar inductor at the time of start-up as:

- Forward|reverse reflection coefficient
- Forward|reverse return loss
- Forward|reverse voltage gain
- Insertion loss

These are shown in Figs. 3.1, 3.2, 3.3, 3.4, 3.5, 3.6, 3.7, 3.8 and 3.9. Simple mathematical formulas relate these small signal performance characteristics to the small signal S (scattering) parameters measured by the SPICE[] simulator using AC (small signal) analysis, e.g., the forward reflection coefficient.

Is $\Gamma_{\text{FORWARD v INOUT}} = 20.0 \log \left(\sqrt{S_{11,REAL}^2 + S_{11,IMAG}^2} \right)$ and so on. Measuring the s parameters is slightly tricky, and the following sequence of instructions must be used:

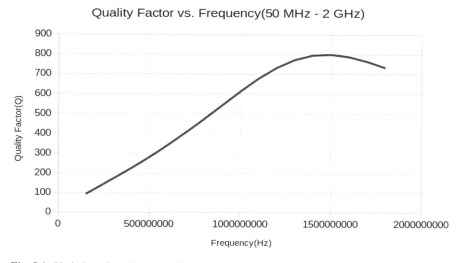

Fig. 3.1 Variation of quality factor Q with frequency for printed circuit board planar rectangular microstrip inductor

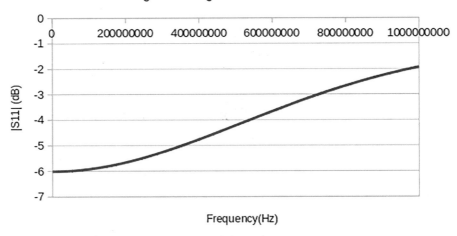

Fig. 3.2 Small signal forward reflection coefficient planar rectangular inductor on printed circuit

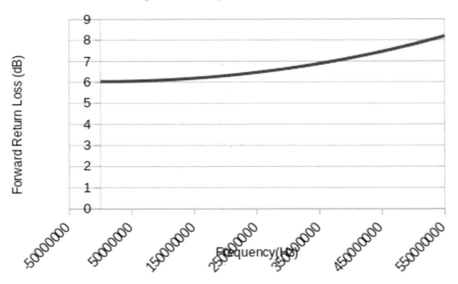

Fig. 3.3 Small signal forward return loss planar rectangular inductor on printed circuit board

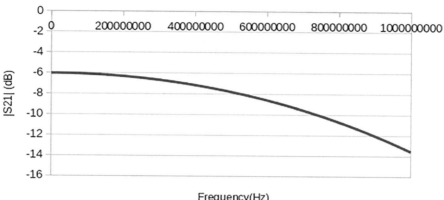

Fig. 3.4 Small signal forward voltage gain planar rectangular inductor on printed circuit board

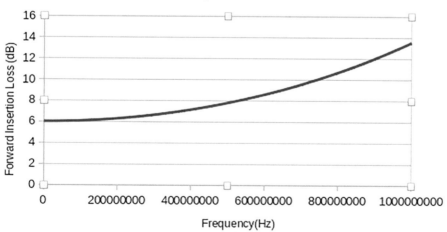

Fig. 3.5 Small signal insertion loss planar rectangular inductor on printed circuit board

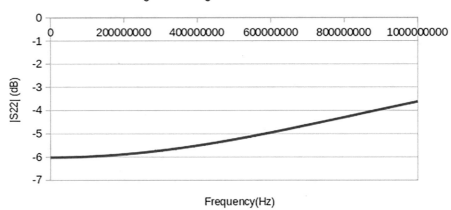

Fig. 3.6 Small signal backward|reverse|output reflection coefficient planar rectangular microstrip inductor

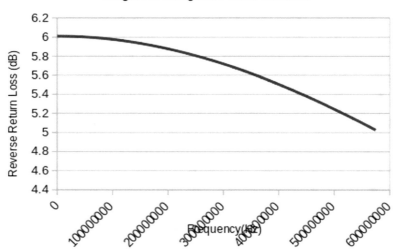

Fig. 3.7 Small signal output return loss planar rectangular inductor on printed circuit board

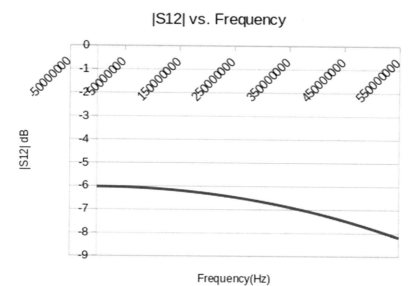

Fig. 3.8 Small signal reverse voltage gain planar inductor on printed circuit board

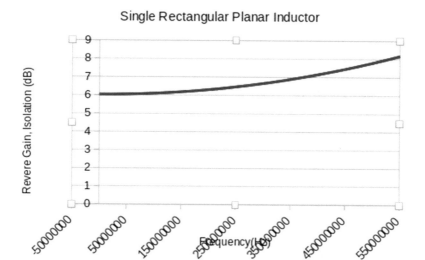

Fig. 3.9 Small signal reverse fain and isolation planar rectangular inductor on printed circuit board

- To measure S11 **XPLINDMSS 2 3 PLINDMSS** and **.PRINT AC V(2)**
- T0 measure S12 **XPLINDMSS 3 2 PLINDMSS** and **.PRINT AC V(3)**
- To measure S22 **XPLINDMSS 3 2 PLINDMSS** and **.PRINT AC V(2)**
- To measure S21 **XPLINDMSS 2 3 PLINDMSS** and **.PRINT AC V(3)**

The supplied C computer language [1] executable ***sparameasdb***, invoked using command line arguments, converts the raw SPICE [2–5] tab separated variable output to meaningful information forward|input reflection coefficient, etc., as shown below: typing **./sparameasdb** at the Linux|MinGW shell command prompt generates hekp information:

```
./sparameasdb
incorrect/insufficient arguments
./sparammeas <input file name>
<output file name> 1|2|3|4
1 -> S11(dB) Input Return Loss(dB)
2 -> S21(dB) Insertion Loss(dB)
3 -> S22(dB) Output Return Loss(dB)
4 -> S12(dB) Reverse Gain(dB) - Reverse Isola
```

3.2 Single Planar Circular Loop Inductor on Printed Circuit Board

The supplied C computer language [1] executable ***planarind*** can also be used to generate the SPICE [2–5] netlist for a planar circular loop inductor using appropriate batch|command line arguments for a circular loop inductor on a printed circuit board as:

```
./planarind C 2 50 4.75 500 7.6 1.5 0.01 0.75 0 0 0.95
mean radius 0.02500 m
calculated inductor 8.611078e-06 H
SPICE netliat planarind.cir
```

The generated SPICE [2–5] netlist is listed below:

```
PLANAR CIRCULAR LOOP INDUCTOR

.PARAMS C=7.629827e-10 FREQ=5.00000e+08 L=8.611078e-06 R=69.3751
.PARAMS FLLIM=2.500000e+06 FHLIM=2.500000e+10 AMPL=5 CT=1.0E-15
RR=50.0 RSMALL=0.001 TS='1.0/(20.0*FREQ)' TSTOP='50.0/FREQ'
TSTRT='1.0/FREQ' TOT=5000
.SUBCKT PLINDCL 1 2
** 1 IN
** 2 OUT
C1 2 0 {C}
L0 1 3 {L}
```

```
R0 3 2 {R}
.ENDS

** TRANSIENT ANALYSIS
R0 1 2 {RSMALL}
VTST0 2 3 DC 0.0 AC 0.0
XPLINDCL 3 0 PLINDCL
VSIG 1 0 DC 0.001 SIN(0 {AMPL} {FREQ} 0 0 0)

** RESOMANCE MEASUREMENT
CT 3 4 {CT}
RS 1 2 {RR}
RL 4 0 {RR}

XPLINDCL 2 3 PLINDCL
VSIG 1 0 DC 0.001 AC {AMPL}

** S PARAMETER MEASUREMENT
RS 1 2 {RR}
RL 3 0 {RR}

** S11 S21
XPLINDCL 2 3 PLINDCL
** S22 S12
XPLINDCL 3 2 PLINDCL
VSIG 1 0 DC 0.001 AC 1.0

.OPTIONS METHOD=GEAR NOPAGE RELTOL=1m
.IC
.TRAN {TS} {TSTOP} {TSTRT} UIC
.PRINT TRAN V(3) I(VTST0)

** S PARAMETER
.AC LIN {TOT} {FLLIM} {FHLIM}
** RESONANCE MEASUREMENT
.PRINT AC V(4)
** S PARAMETER MEASUREMENT
** S11 S22
.PRINT AC V(2)
** S21 S12
.PRINT AC V(3)
.END
```

Appropriate lines in the above netlist are commented out to execute transient and small signal (resonance|s parameter) measurements.

As in the case of the rectangular microstrip inductor on a printed circuit board, a sequence of SPICE [2–5] transient analyses, one for each frequency in the selected band, is executed, to determine the Q factor for the circular loop inductor. The values of the measured RMS voltage across, current through, and frequency-dependent series resistance are listed in tabular form below:

```
1.0E+8,3.531931e+00,6.985728e-04,69.376
2.0E+8,3.531932e+00,3.828248e-04,69.3756
3.0E+8,3.531932e+00,2.730628e-04,69.3755
4.0E+8,3.531932e+00,2.150041e-04,69.3754
5.0E+8,3.531932e+00,1.782580e-04,69.3753
6.0E+8,3.531932e+00,1.526182e-04,69.3753
7.0E+8,3.531932e+00,1.335981e-04,69.3753
8.0E+8,3.531932e+00,1.188791e-04,69.3753
9.0E+8,3.531932e+00,1.071278e-04,69.3752
1.0E+9,3.531932e+00,9.751743e-05,69.3752
1.1E+9,3.531932e+00,8.950564e-05,69.3752
1.2E+9,3.531932e+00,8.272067e-05,69.3752
1.3E+9,3.531932e+00,7.689868e-05,69.3752
1.4E+9,3.531932e+00,7.184694e-05,69.3750
1.5E+9,3.531932e+00,6.742124e-05,69.3750
1.6E+9,3.531932e+00,6.351144e-05,69.3751
1.7E+9,3.531932e+00,6.003193e-05,69.3851
1.8E+9,3.531932e+00,5.691513e-05,69.3751
1.9E+9,3.531932e+00,5.410694e-05,69.3751
2.0E+9,3.531932e+00,5.156356e-05,69.3752
2.1E+9,3.531932e+00,4.924913e-05,69.3751
2.2E+9,3.531932e+00,4.713397e-05,69.3751
2.3E+9,3.531932e+00,4.519337e-05,69.3752
2.4E+9,3.531932e+00,4.340788e-05,69.3751
2.5E+9,3.531932e+00,4.175979e-05,69.3751
```

The C computer language [1] executable *qfactmeas* uses this data to compute the circular loop inductor's quality factor as a function of frequency (Fig. 3.10). The last column in the above data set represents the frequency-dependent series resistance of the circular loop inductor.

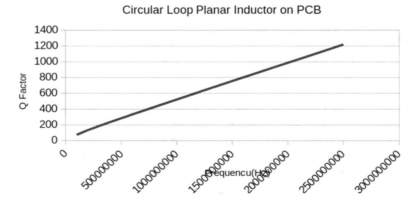

Fig. 3.10 Q factor for single loop inductor on printed circuit board

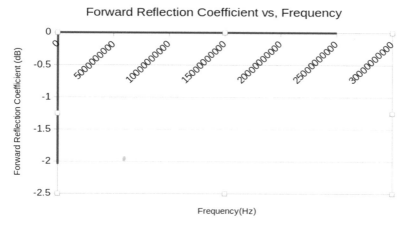

Fig. 3.11 Small signal forward reflection coefficient planar circular loop inductor

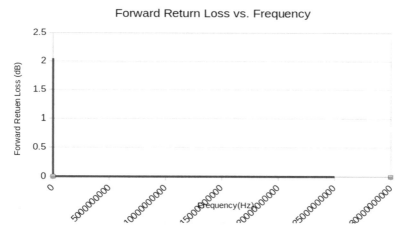

Fig. 3.12 Small signal forward return loss planar circular loop inductor

The start-up performance characteristics, as measured using SPICE[] AC (small signal) analysis, and subsequent processing of the raw data with **sparameasdb**, are shown in Figs. 3.11, 3.12, 3.13, 3.14, 3.15 and 3.16.

3.3   Planar Rectangular Spiral Inductor with Minimum Substrate Loss

The supplied C computer language [1] executable **planarindd** accepts the following set of input as batch|command line argument list and computes the values of the components of the equivalent electrical circuit for a planar square|rectangular spiral

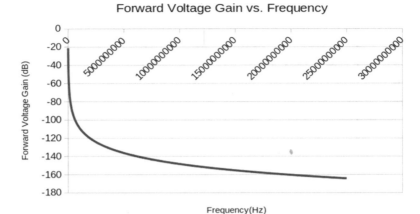

Fig. 3.13 Small signal forward voltage gain circular loop inductor

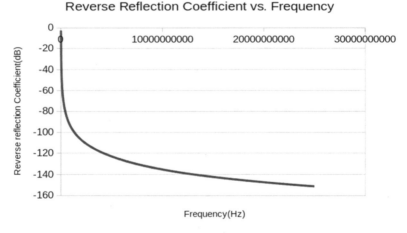

Fig. 3.14 Small signal reverse reflection coefficient circular loop inductor

inductor with ***minimum substrate loss***. Because of the spiral structure of the planar inductor, in addition to the self-inductance of each rectangular segment, ***the anti-parallel (current flowing in opposite directions) and parallel (current flowing in the same direction) mutual inductances of each possible parallel segment combination have to be included in the calculations.*** Mutual inductances arise mainly because of antiparallel current flow in parallel inductor segments (creating at first capacitive coupling between each such parallel inductor pair).

"Minimum loss" indicates that this square|rectangular planar spiral inductor is fabricated on a printed circuit board, so that the unavoidable parasitic capacitive coupling between each inductor segment and the ground plane is the chief energy

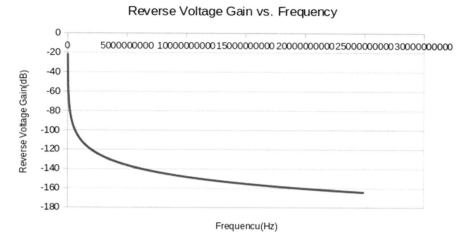

Fig. 3.15 Small signal reverse voltage gain circular loop planar inductor

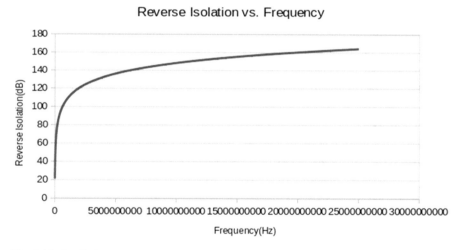

Fig. 3.16 Small signal reverse isolation circular planar loop inductor

loss mechanism. For a planar spiral inductor fabricated inside an integrated circuit, parasitic capacitive coupling between the inductor traces and the silicon oxide layer, as well as the silicon substrate, must be included in the calculations—this will be examined in detail in a subsequent section.

Typing **./planarindd** at the Linux|MinGW shell command prompt generates the following help message, listing the required input parameters.

```
./planarindd
incorrect|insufficient arguments
batch|command line argument mode
./planarindd b|B|c|C
<substrate dielectric constant>
<mid-band|operating frequency(MHz)>
<max. spiral length(mm)>
<max. spiral width(mm)>
<substrate thickness(mm)>
<trace thickness(mm)>
<trace width(mm)>
<trace separation(mm)>
<target inductance(nH)>
<characteristic impedance(Ohm)>
sample batch|command line input
./planarindd B 9 500 20 17 0.75 0.01 1 0.25 100.0 50.0
```

When the supplied sample command line argument input is used, **planarindd** generates the following output, along with the required SPICE [2–5] netlist:

```
./planarindd B 9 100 20 17 0.75 0.01 1 0.25 100.0 50.0
planar spiral inductor segment lengths - m
2.000000e-02
1.600000e-02
1.900000e-02
1.575000e-02
1.775000e-02
1.350000e-02
1.525000e-02
1.100000e-02
1.275000e-02
8.500000e-03
planar spiral inductor segment self-inductances H
7.560825e-09
6.048660e-09
7.182784e-09
5.954150e-09
6.710233e-09
5.103557e-09
5.765129e-09
4.158454e-09
4.820026e-09
3.213351e-09
planar spiral inductor segment mutual inductances H
3.246867e-09
2.597494e-09
3.084524e-09
2.556908e-09
2.881595e-09
2.191635e-09
2.475736e-09
1.785777e-09
```

```
2.069878e-09
1.379919e-09
planar spiral inductor segment end capacitors F
2.132637e-12
1.706109e-12
2.026005e-12
1.679452e-12
1.892715e-12
1.439530e-12
1.626136e-12
1.172950e-12
1.359556e-12
9.063707e-13
planar spiral inductor segment series resistor Ohm
9.926981e-02
7.941585e-02
9.430632e-02
7.817498e-02
8.810196e-02
6.700712e-02
7.569323e-02
5.459840e-02
6.328450e-02
4.218967e-02
Negative valued total inductances
are rejected as physically meaningless
total series resistance 7.420418e-01 Ohm
#inductor      inductance(H)
1      1.080769e-08
2      1.620698e-08
3      2.387679e-08
4      2.930333e-08
5      3.633825e-08
6      4.075184e-08
7      4.680108e-08
8      5.026957e-08
9      5.537370e-08
10     5.789709e-08
Total computed inductance 1.165330e-07 H target 1.000000e-07 H
out of 10 segments use 5
SPICE netlist planarindd.cir
```

The output line **"out of 10 segments use 5"** *indicates that the sum of the first five computed segment values is sufficient to satisfy the design target value of 100.0 nanoHenry(nH), within the tolerance of $\pm$ 95%.* The sum of these inductances is **computed inductance 1.165330e-07 H or 116.535 nH.** The generated SPICE [2–5] netlist is presented below:

RECTANGULAR SPIRAL INDUCTOR - MINIMUM SUBSTRATE LOSS

```
.PARAMS FREQ=1.000000e+08 FLLIM='FREQ/100.0' FHLIM='1.25*FREQ'
AMPL=5 CT=1.0E-15 R=50.0 RSMALL=0.001 TS='1.0/(20.0*FREQ)'
TSTOP='50.0/FREQ' TSTRT='1.0/FREQ' TOT=5000

.SUBCKT PLINDR 1 2
** 1 IN
** 2 OUT
C00 1 0 2.132637e-12
C10 4 0 2.132637e-12
L0 1 3 1.080769e-08
R0 3 4 9.926981e-02
C01 4 0 1.706109e-12
C11 6 0 1.706109e-12
L1 4 5 1.620698e-08
R1 5 6 7.941585e-02
C02 6 0 2.026005e-12
C12 8 0 2.026005e-12
L2 6 7 2.387679e-08
R2 7 8 9.430632e-02
C03 8 0 1.679452e-12
C13 2 0 1.679452e-12
L3 8 9 2.930333e-08
R3 9 2 7.817498e-02
C04 2 0 1.892715e-12
C14 4 0 1.892715e-12
L4 2 11 3.633825e-08
R4 11 4 8.810196e-02
.ENDS

** TRANSIENT ANALYSIS Q FACTOR
R0 1 2 {RSMALL}
VTST0 2 3 DC 0.0 AC 0.0
XL 3 0 PLINDR
VSIG 1 0 DC 0.001
+ SIN(0 {AMPL} {FREQ} 0 0 0)

** RESONANCE MEASUREMENT
CT 3 4 {CT}
RS 1 2 {R}
RL 4 0 {R}
XL 2 3 PLINDR
VSIG 1 0 DC 0.001 AC {AMPL}

** S PARAMETER MEASUREMENT
RS 1 2 {R}
RL 3 0 {R}
** S11 S21
XL 2 3 PLINDR
** S22 S12
XL 3 2 PLINDR
VSIG 1 0 DC 0.001 AC 1.0
```

```
.OPTIONS METHOD=GEAR NOPAGE RELTOL=1m

** TRANSIENT ANALYSIS
.IC
.TRAN {TS} {TSTOP} {TSTRT} UIC
.PRINT TRAN V(3) I(VTST0)

.AC LIN {TOT} {FLLIM} {FHLIM}
** RESONANCE MEASUREMENT
.PRINT AC V(4)

** S PARAMETER S11 S22
.PRINT AC V(2)
** S PARAMETER S21 S12
.PRINT AC V(3)

.END
```

Using SPICE [2–5] transient analysis, with a new SPICE [2–5] netlist each corresponding to a different frequency in the range 100 MHz–2.5 GHz, the RMS current flowing through the square|rectangular planar spiral inductor and the voltage across it, are measured, as listed below. A new SPICE [2–5] netlist is essential for each frequency as the series resistance changes with frequency:

```
1.0E+8,3.530150e+00,1.746568e-01,7.420418e-01
2.0E+8,3.532021e+00,8.014236e-02,8.657747e-01
3.0E+8,3.527890e+00,4.884690e-02,9.681493e-01
4.0E+8,3.531305e+00,1.124165e-01,1.058486e+00
5.0E+8,3.531917e+00,1.956527e-01,1.140559e+00
6.0E+8,3.534891e+00,4.766710e-02,1.216400e+00
7.0E+8,3.530234e+00,8.257575e-02,1.287278e+00
8.0E+8,3.529998e+00,1.013516e-01,1.354060e+00
9.0E+8,3.535477e+00,5.090590e-02,1.417378e+00
1.0E+9,3.531925e+00,8.735490e-02,1.477710e+00
1.1E+9,3.532275e+00,8.034621e-02,1.535432e+00
1.2E+9,3.533225e+00,3.117386e-02,1.590846e+00
1.3E+9,3.532650e+00,2.535887e-02,1.644198e+00
1.3E+9,3.532650e+00,2.535887e-02,1.644198e+00
1.4E+9,3.530778e+00,2.954741e-02,1.695695e+00
1.5E+9,3.535128e+00,3.642447e-02,1.745511e+00
1.6E+9,3.533578e+00,4.386004e-02,1.793795e+00
1.7E+9,3.534806e+00,5.091768e-02,1.840674e+00
1.8E+9,3.532856e+00,5.785971e-02,1.886262e+00
1.9E+9,3.533961e+00,6.449594e-02,1.930655e+00
2.0E+9,3.531123e+00,7.120101e-02,1.973940e+00
2.1E+9,3.530451e+00,7.745228e-02,2.016194e+00
2.2E+9,3.531952e+00,8.348553e-02,2.057484e+00
2.3E+9,3.532929e+00,8.946380e-02,2.097872e+00
2.4E+9,3.535597e+00,9.531586e-02,2.137413e+00
2.5E+9,3.529998e+00,1.013516e-01,2.176157e+00
```

Frequency Dependent Planar Rectangular Spiral Inductor Resistance vs. Frequency

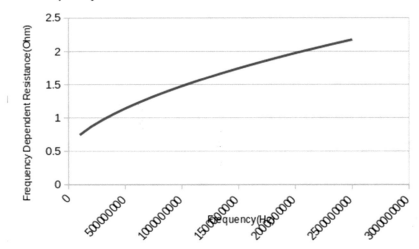

Fig. 3.17 Rectangular planar spiral inductor series resistance minimum substrate loss

In the above listing, for each row:

- The first number is the frequency in Hz
- The second the voltage across the square|rectangular planar spiral inductor in Volts
- The third number the current through the square|rectangular planar spiral inductor in Amperes
- The fourth number is the square|rectangular planar spiral inductor series resistance in Ohms

It is clear that the series resistance is increasing with frequency. Figure 3.17 shows the variation of series resistance with frequency. This resistance increase with increasing frequency is due to skin effect in each inductor segment, and proximity effect in between nearest neighbor adjacent parallel inductor segments.

The Q factor for this square|rectangular minimum loss planar spiral inductor is calculated using the supplied C computer language [1] executable **qfactmeas**, using the above listed data as input. The variation of Q factor with frequency is shown in Fig. 3.18.

The RMS current through the minimum loss substrate planar spiral inductor also varies, as frequency-dependent impedance—consisting of the frequency-dependent series resistance—varies (skin, proximity effects) along with the inductor reactance (directly proportional to the frequency) changes. So for constant RMS voltage drop across the planar spiral inductor, the impedance has to change for changing RMS current through it. The RMS frequency-dependent current through the planar spiral inductor (on minimum loss substrate) is shown in Fig. 3.19.

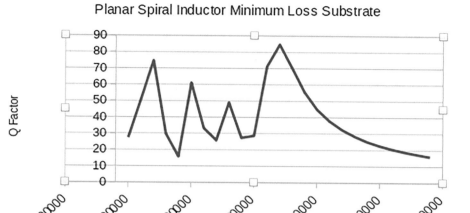

Fig. 3.18   Q factor vs. frequency for planar squarelrectangular inductor with minimum substrate loss

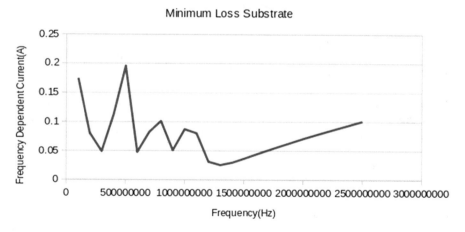

Fig. 3.19   Frequency-dependent current through planar spiral inductor on minimum loss substrate

These SPICE [2–5] transient analysis results show that when current through the minimum loss substrate planar spiral inductor peaks, the Q factor falls and vice versa. As the series resistance varies with frequency as a consequence of skin and proximity effects, standard SPICE [2–5] small signal (AC) analysis would not provide any meaningful results, if the small signal analysis is done over a large (several hundred MHz) frequency range The result of such a restricted small signal analysis is shown in Fig. 3.20.

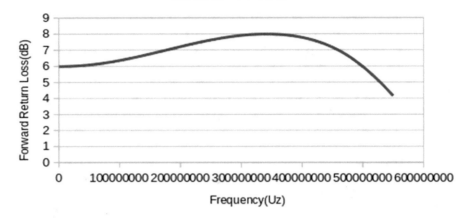

Fig. 3.20 Restricted narrow band small signal forward return loss planar spiral inductor on minimum loss substrate

3.4 Integrated (On-Chip) Planar Square|Rectangular Spiral Inductor on Fully Lossy Substrate: Method A

The supplied C computer language [1] executable ***planarindgh*** exploits the Greenhouse formalism and some modern refinements [6] to compute the values if the equivalent circuit for a square|rectangular|hexagonal|octagonal spiral inductor fabricated inside an integrated circuit. ***This means that the substrate is fully lossy, i.e., parasitic oxide, substrate capacitances, and parasitic substrate resistors are associated with each segment of the planar spiral inductor.*** Typing ./planarindgh at the Linux|MinGW shell command prompt generates the following help information:

```
./planarindgh
incorrect/insufficient arguments
interactive mode
./spiralind_1 i|I 1|2|3
batch/command line argument mode
./planarindgh b|B|c|C 1|2|3
<trace width(mm)>
<trace separation(mm)>
<oxide thickness(mm)>
<substrate thickness(mm)>
<trace thickness(mm)>
<target inductance value(nH)>
<operating frequency(MHz)>
<inductor maximum length(mm)>
<inductor maximum width(mm)>
```

```
<maximum outer diameter (mm) >
maximum length, width non-zero ONLY
<under pass depth (mm)
<tolerance e.g., 0.95>
tolerance determines how close the
computed inductance (self + mutual)
value is to the specified target value
for square, rectangle spiral inductor
maximum spiral outer diameter non-zero
ONLY for hexagon, octagon spiral inductor
1 -> rectangle|square spiral inductor
2 -> hexagon spiral inductor
3 -> octagon spiral inductor
sample command line input square|rectangle spiral inductor
./planarindgh b 1 1.25 0.5 0.45 1.75 0.001 30 200 60 70 0 0.5 0.95
sample command line input hexagon spiral inductor
./planarindgh C 2 1.25 0.5 0.45 1.75 0.001 150 100 0 0 80 0.5 0.95
Only parallel inductor segment pairs
contribute to mutual inductance
to ensure accurate sequence of input parameters, it is best to execute
the program using command line arguments
```

Using the supplied sample command line argument line input for a square|
rectangular planar spiral inductor operating at a frequency of 100 MHz, the program
generates the following output:

```
./planarindgh b 1 1.25 0.5 0.45 1.75 0.001 30 100 60 70 0 0.5 0.95
Tol 1 0.950000
target 3.000000e-08 H   computed maximum 2.756018e-08 H
total number of traces 16
total inductor length 0.808500 mm
skin depth 6.526720e-06 m
frequency dependent series resistance 22.586172 Ohm
substrate resistance 1108.225108 Ohm
series capacitance 1.106750e-13 F
substrate capacitance 1.890461e-08 F
oxide capacitance 1.080263e-11 F
Length(m) self-inductance(H mutual inductance(H) total inductance(H)
0 6.000000e-02 1.717523e-09 0.000000e+00 1.717523e-09
1 7.000000e-02 2.003777e-09 0.000000e+00 3.721300e-09
2 5.825000e-02 1.667429e-09 0.000000e+00 5.388729e-09
3 6.825000e-02 1.953683e-09 0.000000e+00 7.342412e-09
4 5.475000e-02 1.567240e-09 4.565003e-10 9.366152e-09
5 5.475000e-02 1.567240e-09 4.565003e-10 1.138989e-08
6 5.125000e-02 1.467051e-09 4.204431e-10 1.327739e-08
7 5.125000e-02 1.467051e-09 4.204431e-10 1.516488e-08
8 4.775000e-02 1.366862e-09 3.848645e-10 1.691661e-08
9 4.775000e-02 1.366862e-09 3.848645e-10 1.866833e-08
10 4.425000e-02 1.266673e-09 3.497999e-10 2.028481e-08
11 4.425000e-02 1.266673e-09 3.497999e-10 2.190128e-08
12 4.075000e-02 1.166484e-09 3.152899e-10 2.338305e-08
13 4.075000e-02 1.166484e-09 3.152899e-10 2.486483e-08
```

```
14 3.725000e-02 1.066296e-09 2.813824e-10 2.621251e-08
15 3.725000e-02 1.066296e-09 2.813824e-10 2.756018e-08
SPICE netlist planarindgh.cir
```

The generated SPICE [2–5] netlist is listed below:

SQUARE | RECTANGULAR PLANAR INDUCTOR LUMPED MODEL

```
.PARAMS FREQ=1.000000e+08 FLLIM='0.01*FREQ' FHLIM='1.75*FREQ'
+ RSMALL=0.001 TS='1.0/(20.0*FREQ)'
+ TSTOP='50.0/FREQ' TSTRT='1.0/FREQ' TOT=5000
.PARAMS COX=1.080263e-11 CSER=1.106750e-13
.PARAMS CSUB=9.452305e-09 LS=2.756018e-08
.PARAMS RSER=22.586172 RSUB=554.112554 R=50.0 C=1.0E-8 AMPL=10
.PARAMS RSS=50.0 RSL=50.0
.PARAMS CL=10.0 CTEST=1.0E-15

.SUBCKT PLANARSPIRALIND 1 2
** 1 IN
** 2 OUT
CSER 1 2 {CSER}
COX0 1 4 {COX}
COX1 2 5 {COX}
CSUB0 4 0 {CSUB}
CSUB1 5 0 {CSUB}
L 1 3 {LS}
RSER 3 2 {RSER}
RSUB0 4 0 {RSUB}
RSUB1 5 0 {RSUB}
.ENDS

.SUBCKT PLANARINDXFRMR 1 2 3 4
** 1 IN 1
** 2 OUT 1
** 3 IN 2
** 4 OUT 2
COX11 1 9 {COX}
COX12 2 10 {COX}
COX21 3 7 {COX}
COX22 4 8 {COX}
CS111 9 0 {CSUB}
CSI12 10 0 {CSUB}
CS121 7 0 {CSUB}
CSI32 8 0 {CSUB}
CSER1 1 2 {CSER}
CSER2 3 4 {CSER}
L1 1 5 {LS}
L2 3 6 {LS}
RSER1 2 5 {RSER}
RSER2 4 6 {RSER}
RSI11 9 0 {RSUB}
RSI12 10 0 {RSUB}
```

```
RSI21 7 0 {RSUB}
RSI22 8 0 {RSUB}
k0 L1 L2 0.99
.ENDS


** COMMENT OUT AC OR TRANSIENT ANALYSIS AS NEEDED
** TRANSIENT ANALYSIS Q FACTOR
R0 1 2 {RSMALL}
VTST0 2 3 DC 00 AC 0.0
XPLI 3 0 PLANARSPIRALIND
VSIG 1 0 DC 0.001
+ SIN(0 {AMPL} {FREQ} 0 0 0)


** AC SMALL SIGNAL RESONANCE
CT 2 3 {CTEST}
RS 1 2 {R}
RL 4 0 {R}
XPLI 3 4 PLANARSPIRALIND
VSIG 1 0 DC 0.001 AC {AMPL}


** AC SMALL SIGNAL S PARAMETER
RS 1 2 {R}
RL 3 0 {R}
** S11 S21
XPLI 2 3 PLANARSPIRALIND
** S22 S12
XPLI 3 2 PLANARSPIRALIND
VSIG 1 0 DC 0.001 AC 1.0


.OPTIONS METHOD=GEAR NOPAGE RELTOL=1m

** TRANSIENT ANALYSIS Q FACTOR
.IC
.TRAN {TS} {TSTOP} UIC
.PRINT TRAN V(3) I(VTST0)

.AC LIN {TOT} {FLLIM} {FHLIM}
** AC SMALL SIGNAL RESONANCE
.PRINT AC V(4)

** S PARAMETER S11 S22
.PRINT AC V(2)
** S PARAMETER S21 S12
.PRINT AC V(3)


.END
```

The values of RMS frequency-dependent current, frequency-dependent resistance, and Q factor (listed in tabular format next) for each frequency in the range 100 MHz–2 GHz are listed below. These RMS values were computed using the output of SPICE [2–5] transient analysis performed with a separate SPICE [2–5] netlist, each corresponding to each selected frequency in the abovementioned frequency band. This has been done to account for the frequency-dependent total series resistance of the planar spiral inductor (on lossy substrate). Figures 3.21, 3.22 and 3.23 show respectively the frequency-dependent resistance, frequency-dependent current, and Q factor for the square|rectangular planar spiral inductor on the lossy substrate:

```
1.0E+8,7.008614e+00,2.161081e-01,22.586172
2.0E+8,7.018826e+00,1.005955e-01,22.952213
3.0E+8,7.018787e+00,6.843462e-02,23.238010
4.0E+8,7.018804e+00,1.191491e-01,23.482262
5.0E+8,7.018813e+00,1.807212e-01,23.699957
6.0E+8,7.018817e+00,2.407043e-01,23.898785
7.0E+8,7.018820e+00,2.986020e-01,24.083311
8.0E+8,7.018820e+00,3.548726e-01,24.256512
9.0E+8,7.018821e+00,4.099392e-01,24.420453
1.0E+9,7.018821e+00,4.641105e-01,24.576641
1.1E+9,7.018822e+00,5.176048e-01,24.726210
1.2E+9,7.018821e+00,5.705772e-01,24.870044
1.3E+9,7.018820e+00,6.231403e-01,25.008844
1.4E+9,7.018820e+00,6.753766e-01,25.143180
1.5E+9,7.018820e+00,7.273484e-01,25.273520
1.6E+9,7.018818e+00,7.791034e-01,25.400257
1.7E+9,7.018817e+00,8.306784e-01,25.523722
1.8E+9,7.018817e+00,8.821021e-01,25.644198
1.9E+9,7.018816e+00,9.333976e-01,25.761931
2.0E+9,7.018814e+00,9.845839e-01,25.877132
```

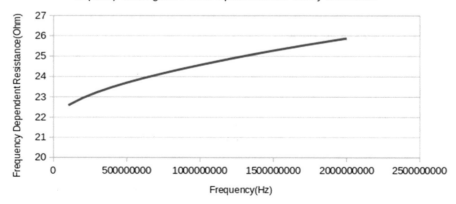

Fig. 3.21 Frequency-dependent resistance square|rectangular planar spiral inductor on lossy substrate

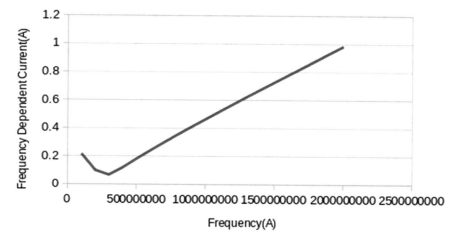

Fig. 3.22 Frequency-dependent current planar spiral inductor lossy substrate

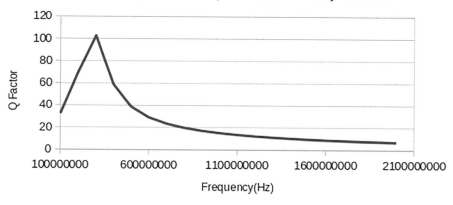

Fig. 3.23 Q factor planar spiral inductor lossy substrate

Clearly, both RMS current through the planar spiral inductor and its series resistance increase with frequency, and so with constant RMS voltage across the planar spiral inductor, its impedance decreases with frequency. As the inductive reactance is directly proportional to the frequency, this implies that the impedance is decreasing with frequency. This is a direct consequence of the parasitic oxide and substrate capacitances. The value of the Q factor peaks at approximately 300 MHz, and then drops off—negative Q factors do not have any physical meaning. Effectively, the inductor starts behaving as a capacitor.

3.5 Integrated (On-Chip) Hexagonal Planer Spiral Inductor on Fully Lossy Substrate: Method A

The C computer language executable [1] ***planarindgh*** can be used to generate the SPICE [2–5] netlist for a planar hexagonal spiral inductor on a lossy substrate (in addition to a planar square|rectangular spiral inductor) simply by specifying the type (1 → square|rectangular, 2 → hexagon). The results generated are listed below:

```
./planarindgh C 2 1.25 0.5 0.45 1.75 0.001 100 100 0 0 80 0.5 0.95
total number of traces 8
self-inductance total 70.532808 (nH)
target 1.000000e-07 H
computed maximum 1.666396e-07 H
total number of traces 8
total inductor length 19.997111 mm
skin depth 6.526720e-06 m
frequency dependent series resistance 558.637202 Ohm
substrate resistance 44.806473 Ohm
series capacitance 3.689167e-14 F
substrate capacitance 4.675790e-07 F
oxide capacitance 2.671880e-10 F
Length(mm) self-inductance(nH) mutual inductance(nH)
0 4.619 18.239 0.000 18.239
1 4.013 15.297 0.000 33.537
2 3.408 12.447 0.000 45.984
3 2.802 9.705 0.000 55.689
4 2.197 7.095 0.000 62.785
5 1.591 4.654 0.000 67.439
6 0.986 2.449 0.000 69.888
7 0.380 0.644 0.000 70.533
SPICE netlist planarindgh.cir
```

The generated SPICE [2–5] netlist are listed next:

```
HEXAGONAL PLANAR INDUCTOR LUMPED MODEL

.PARAMS FREQ=1.000000e+08 FLLIM='0.01*FREQ' FHLIM='1.75*FREQ'
+ RSMALL=0.001 TS='1.0/(20.0*FREQ)'
+ TSTOP='50.0/FREQ' TSTRT='1.0/FREQ' TOT=5000
```

```
.PARAMS COX=2.671880e-10 CSER=3.689167e-14
.PARAMS CSUB=2.337895e-07 LS=1.666396e-07
.PARAMS RSER=558.637202 RSUB=22.403236 R=50.0 C=1.0E-8 AMPL=10
.PARAMS RSS=50.0 RSL=50.0
.PARAMS CL=10.0 CTEST=1.0E-15

.SUBCKT PLANARSPIRALIND 1 2
** 1 IN
** 2 OUT
CSER 1 2 {CSER}
COX0 1 4 {COX}
COX1 2 5 {COX}
CSUB0 4 0 {CSUB}
CSUB1 5 0 {CSUB}
L 1 3 {LS}
RSER 3 2 {RSER}
RSUB0 4 0 {RSUB}
RSUB1 5 0 {RSUB}

.ENDS

.SUBCKT PLANARINDXFRMR 1 2 3 4
** 1 IN 1
** 2 OUT 1
** 3 IN 2
** 4 OUT 2
COX11 1 9 {COX}
COX12 2 10 {COX}
COX21 3 7 {COX}
COX22 4 8 {COX}
CS111 9 0 {CSUB}
CSI12 10 0 {CSUB}
CS121 7 0 {CSUB}
CSI32 8 0 {CSUB}
CSER1 1 2 {CSER}
CSER2 3 4 {CSER}
L1 1 5 {LS}
L2 3 6 {LS}
RSER1 2 5 {RSER}
RSER2 4 6 {RSER}
RSI11 9 0 {RSUB}
RSI12 10 0 {RSUB}
RSI21 7 0 {RSUB}
RSI22 8 0 {RSUB}
k0 L1 L2 0.99
.ENDS

** COMMENT OUT AC OR TRANSIENT ANALYSIS AS NEEDED
** TRANSIENT ANALYSIS Q FACTOR
R0 1 2 {RSMALL}
VTST0 2 3 DC 00 AC 0.0
XPLI 3 0 PLANARSPIRALIND
```

```
VSIG 1 0 DC 0.001
+ SIN(0 {AMPL} {FREQ} 0 0 0)


** AC SMALL SIGNAL RESONANCE
CT 2 3 {CTEST}
RS 1 2 {R}
RL 4 0 {R}
XPLI 3 4 PLANARSPIRALIND
VSIG 1 0 DC 0.001 AC {AMPL}


** AC SMALL SIGNAL S PARAMETER
RS 1 2 {R}
RL 3 0 {R}
** S11 S21
XPLI 2 3 PLANARSPIRALIND
** S22 S12
XPLI 3 2 PLANARSPIRALIND
VSIG 1 0 DC 0.001 AC 1.0


.OPTIONS METHOD=GEAR NOPAGE RELTOL=1m


** TRANSIENT ANALYSIS Q FACTOR
.IC
.TRAN {TS} {TSTOP} UIC.PRINT TRAN V(3) I(VTST0)


.AC LIN {TOT} {FLLIM} {FHLIM}
** AC SMALL SIGNAL RESONANCE
.PRINT AC V(4)


** S PARAMETER S11 S22
.PRINT AC V(2)
** S PARAMETER S21 S12
.PRINT AC V(3)


.END
```

SPICE [2–5] transient analysis is performed to determine the RMS current through, the RMS voltage across the hexagonal planar spiral inductor on a lossy substrate. The RMS values for the current through and voltage across the inductor, for each frequency (in predefined band in this case 100 MHz–2 GHz), are used to estimate the inductor's quality factor (Q). These RMS current values are listed below:

```
1.0E+8,7.012969e+00,1.227689e+00,558.637202
2.0E+8,7.015977e+00,2.454457e+00,567.690721
3.0E+8,7.015854e+00,3.681971e+00,574.759510
4.0E+8,7.015709e+00,4.909633e+00,580.800729
5.0E+8,7.015523e+00,6.137299e+00,586.185128
```

```
6.0E+8,7.015378e+00,7.364784e+00,591.102849
7.0E+8,7.015430e+00,8.591798e+00,595.666846
8.0E+8,7.015491e+00,9.818457e+00,599.950716
9.0E+8,7.009497e+00,1.105322e+01,604.005578
1.0E+9,7.015408e+00,1.227113e+01,607.868653
```

It is noticed that the current increases monotonically, indicating that the impedance decreases correspondingly. The frequency-dependent series resistance also increases monotonically. In combination, these guarantee that this hexagonal planar spiral inductor on a lossy substrate cannot have a Q value in this frequency range. The frequency-dependent current through the inductor and the frequency-dependent series resistance are shown in Figs. 3.24 and 3.25.

Clearly, the losses through the substrate increase with frequency. The reason behind this is the oxide and substrate capacitances, whose reactances decrease as frequency increases, thereby turning into short circuits and draining signal energy to ground. The supplied C computer language [1] executable ***planarindgh*** can also generate the SPICE [2–5] netlist for an octagonal planar spiral inductor on a lossy substrate. This is left as an exercise for the reader.

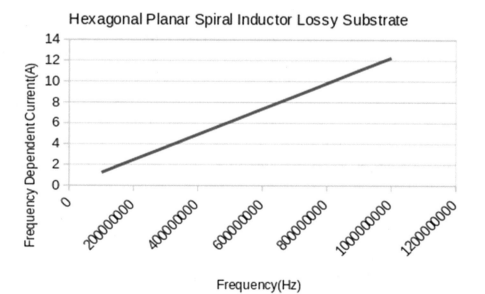

Fig. 3.24 Frequency-dependent current hexagonal planar spiral inductor lossy substrate

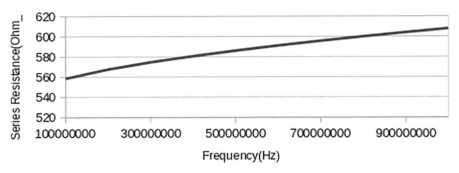

Fig. 3.25 Frequency-dependent series resistance hexagonal planar spiral inductorlossy substrate

3.6   Square|Rectangular Planar Spiral Inductor on Fully Lossy Substrate: Method B

The supplied C computer language [1] executable ***planarindls1*** uses a straightforward implementation of the Greenhouse formalism for a planar spiral inductor on a lossy substrate, without any refinements [6] (as in the previous two analysis|design examples). Typing ./planarindls1 at the Linux|MinGW shell command prompt generates the following help information:

```
./planarindls1
incorrect|insufficient arguments
batch|command line argument mode
./planarindls1 b|B|c|C
<substrate dielectric constant>
<mid-band|operating frequency(MHz)>
<max. spiral length(mm)>
<max. spiral width(mm)>
<substrate thickness(mm)
<trace thickness(mm)
<trace width(mm)
<trace separation(mm)
<trace - underpass separation(mm)>
<oxide thickness(mm)>
<target inductance(nH)<
<characteristic impedance(Ohm)>
<tolerance e.g. 0.95>
tolerance determines how close the
```

```
computed inductance(self + mutual)
value needs to be to the design target value
sample batch|command line input
./planarindls1 B 9 500 20 17 0.75 0.01 1 0.25 0.01 0.2 100.0 50.0 0.95
```

Using the supplied sample command line argument input at an operating frequency of 100 MHz generates the following output:

```
./planarindls1 B 9 100 20 17 0.75 0.01 1 0.25 0.01 0.2 100.0 50.0 0.95
Dielectric constant 9.000000
frequency(MHz) 100.000000
max. length 20.000000 mm
max. width 17.000000 mm
substrate thickness 0.750000 mm
trace thickness 0.010000 mm
trace width 1.000000 mm
trace separation 0.250000 mm
underpass depth 0.010000 mm
oxide thickness 0.200000 mm
target inductance 100.000000 nH
characteristic impedance 50.000000 Ohm
tolerance 0.950000
initialized arrays
computed trace lengths
computed self-inductances
computed mutual inductances
computed total inductances
computed capacitances
computed resistances
inductor segment lengths - m
2.000000e-02
1.600000e-02
1.900000e-02
1.575000e-02
1.775000e-02
1.350000e-02
1.525000e-02
1.100000e-02
1.275000e-02
8.500000e-03
self-inductance H
7.371610e-09
5.897288e-09
7.003030e-09
5.805143e-09
6.542304e-09
4.975837e-09
5.620853e-09
4.054386e-09
4.699401e-09
3.132934e-09
mutual inductances H
```

```
0.000000e+00
0.000000e+00
5.091417e-10
8.341095e-10
8.341095e-10
1.252350e-09
1.563071e-09
2.568579e-09
7.877704e-10
1.087284e-09
Negative valued total inductances
are rejected as physically meaningless
#inductor(s)      inductance (Henry)
1     7.371610e-09
2     1.326890e-08
3     2.078107e-08
4     2.691118e-08
5     3.345348e-08
6     3.884756e-08
7     4.477914e-08
8     4.983903e-08
9     5.275762e-08
10    5.619007e-08
Total computed inductance 3.441997e-07 H
Total computed inductor length 0.1495 m 2.500 turns
microstrip capacitor 1.650078e-11 F
oxide capacitor 1.081424e-11 F
substrate capacitor 7.956855e-12 F
series resistance 0.5028 Ohm silicon resistance 3210.7023 Ohm
SPICE netlist planarindlossy.cir
```

The contents of the generated SPICE [2–5] are listed below:

```
SQUARE PLANAR INDUCTOR LOSSY SUBSTRATE

.PARAMS CP=1.650078e-11 COX=1.081424e-11 CSI=7.956855e-12
LT=3.441997e-07
.PARAMS RSER=0.5028 RSI=3210.7023 AMPL=2 CT=1.0E-15 R=50.0 TOT=50000
.PARAMS FREQ=1.000000e+08 FLLIM='0.01*FREQ' FHLIM='1.25*FREQ'
+ RSMALL=0.001 TS='1.0/(20.0*FREQ)' TSTOP='20.0/FREQ'
+ TSTRT='1.0/FREQ' TOT=100000

.SUBCKT PLINDLSY 1 2
** 1 IN
** 2 OUT
CP 1 2 {CP}
COX0 1 4 {COX}
COX1 2 5 {COX}
CSI0 4 0 {CSI}
CSI1 5 0 {CSI}
LT 1 3 {LT}
RSER 3 2 {RSER}
```

```
RSI0 4 0 {RSI}
RSI1 5 0 {RSI}
.ENDS

** TRANSIENT ANALYSIS Q FACTOR
R0 1 2 {RSMALL}
VTST0 2 3 DC 0.0 AC 0.0

XLPLINDLSY 3 0 PLINDLSY
VSIG 1 0 DC 0.001
+ SIN(0 {AMPL} {FREQ} 0 0 0)

** AC SMALL SIGNAL RESONANCE
CT 3 4 {CT}
RS 1 2 {R}
RL 4 0 {R}
XLPLINDLSY 2 3 PLINDLSY

VSIG 1 0 DC 0.001 AC {AMPL}

** AC SMALL SIGNAL S PARAMETER
RS 1 2 {R}
RL 3 0 {R}
** S11 S21
XLPLINDLSY 2 3 PLINDLSY
** S22 S12
XLPLINDLSY 3 2 PLINDLSY
VSIG 1 0 DC 0.001 AC {AMPL}

.OPTIONS METHOD=GEAR NOPAGE RELTOL=1m

** TRANSIENT ANALYSIS Q FACTOR
.IC
.TRAN {TS} {TSTOP} {TSTRT} UIC
.PRINT TRAN V(3) I(VTST0)

.AC LIN {TOT} {FLLIM} {FHLIM}
** AC SMALL SIGNAL RESONANCE
.PRINT AC V(4)
* AC SMALL S PARAMETER S11 S22
.PRINT AC V(2)
** AC SMALL S PARAMETER S21 S12
.PRINT AC V(3)

.END
```

SPICE [2–5] transient analysis is used to determine the RMS current and RMS voltage at each selected frequency in the band 100.0 MHz–3.0 GHz. This is achieved by generating a fresh SPICE [2–5] netlist for each selected frequency and then using

SPICE [2–5] transient analysis with each generated netlist. The supplied C computer language [1] executable *rmscalc* is used to compute the RMS current through and voltage across the planar spiral inductor on the lossy substrate. This is essential, as the inductor's series resistance is frequency dependent. The values for the frequencies, and the corresponding RMS current, RMS voltage, and frequency-dependent resistance, are listed in tabular format below:

```
1.0E+8,1.412772e+00,1.452818e-02,0.5028
2.0E+8,1.412772e+00,3.570191e-02,0.5030
3.0E+8,1.412773e+00,5.594982e-02,0.5032
4.0E+8,1.412772e+00,7.577206e-02,0.5033
5.0E+8,1.412772e+00,9.540583e-02,0.5035
6.0E+8,1.412772e+00,1.149414e-01,0.5036
7.0E+8,1.412772e+00,1.344196e-01,0.5037
8.0E+8,1.412772e+00,1.538616e-01,0.5038
9.0E+8,1.412772e+00,1.732792e-01,0.5039
1.0E+9,1.412770e+00,1.926780e-01,0.5039
1.1E+9,1.402104e+00,2.104618e-01,0.5040
1.2E+9,1.412771e+00,2.314393e-01,0.5041
1.3E+9,1.412771e+00,2.508083e-01,0.5042
1.4E+9,1.412770e+00,2.701720e-01,0.5042
1.5E+9,1.412771e+00,2.895486e-01,0.5043
1.6E+9,1.412770e+00,3.089041e-01,0.5044
1.7E+9,1.412770e+00,3.282564e-01,0.5044
1.8E+9,1.412770e+00,3.476060e-01,0.5045
1.9E+9,1.412769e+00,3.669534e-01,0.5046
2.0E+9,1.412770e+00,3.862992e-01,0.5046
2.1E+9,1.412769e+00,4.056369e-01,0.5047
2.2E+9,1.412769e+00,4.249836e-01,0.5048
2.3E+9,1.412769e+00,4.443247e-01,0.5049
2.4E+9,1.412769e+00,4.636647e-01,0.5050
2.5E+9,1.412769e+00,4.830038e-01,0.5051
2.6E+9,1.412667e+00,5.018641e-01,0.5049
2.7E+9,1.412768e+00,5.216802e-01,0.5050
2.8E+9,1.413488e+00,5.412942e-01,0.5051
2.9E+9,1.412769e+00,5.603890e-01,0.5050
3.0E+9,1.412768e+00,5.797255e-01,0.5051
```

The frequency-dependent current, resistance, and computed Q factor values are shown in Figs. 3.26, 3.27 and 3.28 respectively. These results demonstrate that the supposed improvements [6] do not have any significant effect because for both cases (Greenhouse only, Greenhouse and [6]), the current through the planar spiral inductor (on lossy substrate) current through it increases rapidly with the frequency of the applied input signal. Consequently, the quality factor (Q) decreases.

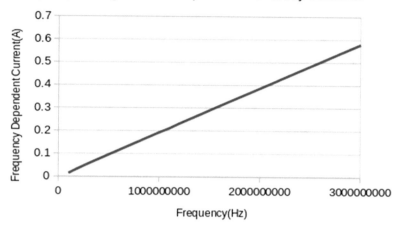

Fig. 3.26 Frequency-dependent current planar spiral inductor lossy substrate

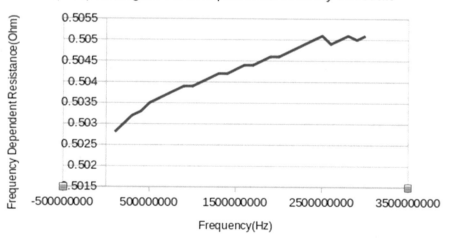

Fig. 3.27 Frequency-dependent resistance planar spiral inductor with lossy substrate

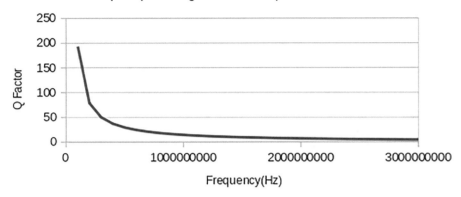

Q Factor vs. Frequency

Square|Rectangular Planar Spiral Inductor

Fig. 3.28 Q factor planar spiral inductor lossy substrate

3.7 Integrated (On-Chip) Square|Rectangular Planar Spiral Inductor on Lossy Substrate Mohan: Wheeler Current Sheet Approach

The Mohan Wheeler scheme to compute the total inductance of a planar spiral inductor on a lossy substrate differs from the Greenhouse scheme, as this new procedure does not explicitly compute the mutual inductance between parallel inductor segments. Given a target total inductance value, along with other key input parameters (trace width, trace separation, etc.), the trace length is adjusted so that the computed total inductance value matches the target value, within applicable tolerances. The supplied C computer language [1] executable *planarindmw* accepts a set of input parameter and computes the values for the equivalent electrical circuit for the planar spiral inductor (on lossy substrate) using the Mohan Wheeler scheme. The output generated by *planarindmw* is text file with the computed electrical circuit component values formatted as a SPICE [2–5] netlist. Typing *./planarindmw* at the Linux|MinGW shell command prompt generates the following help information:

```
./planarindmw
incorrect|insufficient parameters
interactive mode
./planarindmw i|I 1|2|3
batch/command line argument mode
./planarindmw b|B|c|C 1|2|3
<inner diameter(mm)>
<outer diameter(mm)>
```

```
<operating frequency(MHz)>
<target inductance value(nH)>
<trace separation(mm)>
<substrate height(mm)>
<oxide height(mm)>
<magnetic material height(mm)>
<buried trace depth(mm)>
sample batch/command line input planar square|rectangle inductor
./planarindmw b 1 20 45 350 25 0.5 2 0.5 1 0.25
sample batch/command line input planar hexagon inductor
./planarindmw C 2 15 35 500 25 0.5 1.75 0.5 1 0.25
1 -> square|rectangle planar spiral
2 -> hexagon planar spiral
3 -> octagon planar spiral
```

Using the supplied sample command line input for a square|rectangular planar
spiral inductor, an operating frequency of 100 MHz generates the output:

```
./planarindmw b 1 20 45 100 25 0.5 2 0.5 1 0.25
square|rectangle planar spiral inductor
computed physical parameters
number of turns Wheeler 0.732491
number of turns Mohan 0.737859
skin depth 1.149126e-05 m
trace thickness 1.723689e-05 m
trace width 2.872815e-05 m
average conductor length 1.316787e-01 m
final trace separation 5.000000e-04 m
computed small signal model components
substrate capacitance 1.099521e-10 F
series capacitance 0.000000e+00 F
underpass capacitance 1 9.627988e-17 F
underpass capacitance 2 1.771986e-15 F
oxide capacitance 1.506194e-16 F
series resistance 4.520617 Ohm
substrate resistance 676732.087024 Ohm
magnetic layer resistance 0.000005 Ohm
buried trace resistance 0.000917 Ohm
SPICE netlist planarindmw.cir
```

The contents of the generated SPICE [2–5] text netlist are:

```
SPIRAL INDUCTOR WHEELER MOHAN

.PARAMS L=2.500000e-08 CSUB=5.497607e-11 CSER=0.000000e+00
.PARAMS CAPU1=9.627988e-17 CAPU2=1.771986e-15 COX=1.506194e-16
.PARAMS RSER=4.520617e+00 RSUB=3.383660e+05 RMAG=3.172182e-06
RBURCOND=0.000917
.PARAMS FREQ=1.000000e+08 FLLIM='0.01*FREQ'
+ FHLIM='1.75*FREQ' CT=1.0e-15 AMPL=10
+ R=50.0 RSMALL=0.001 TS='1.0/(20.0*FREQ)'
```

```
+ TSTOP=10.0us TSTRT=1.0us TOT=100000
.PARAMS RSS=50.0 RSL=50.0 CL=10.0

.SUBCKT PLANARIND2 1 2
** 1 IN
** 2 OUT
CSER 1 3 {CSER}
COX1 1 6 {COX}
COX2 3 4 {COX}
CSUB1 7 0 {CSUB}
CSUB2 5 0 {CSUB}
CU1 3 2 {CAPU1}
CU2 2 4 {CAPU2}
L0 1 8 {L}
RSER 8 3 {RSER}

RMAG1 6 7 {RMAG}
RMAG2 4 5 {RMAG}
RSUB1 7 0 {RSUB}
RSUB2 5 0 {RSUB}
RBC 3 2 {RBURCOND}
.ENDS

** COMMENT OUT APPROPRIATE SECTIONS TO EXECUTE

** TRANSIENT ANALYSIS Q FACTOR
R0 1 2 {RSMALL}
VTST0 2 3 DC 0.0 AC 0.0
XSIMW 3 0 PLANARIND2
VSIG 1 0 DC 0.001
+ SIN(0 {AMPL} {FREQ} 0 0 0)

** AC SMALL SIGNAL ANALYSIS RESONANCE
RS 1 2 {R}
CT 3 4 {CT}
RL 4 0 {R}
XL0 2 3 PLANARIND2
VSIG 1 0 DC 0.001 AC {AMPL}

** AC SMALL SIGNAL ANALYSIS S PARAMETER
RS 1 2 {R}
RL 3 0 {R}
** S11 S21
XL0 2 3 PLANARIND2
** S22 S12
XL0 3 2 PLANARIND2
VSIG 1 0 DC 0.001 AC 1.0

.OPTIONS METHOD=GEAR NOPAGE RELTOL=1m MINBREAK=4ps

** TRANSIENT ANALYSIS
.IC
```

```
.TRAN {TS} {TSTOP} {TSTRT} UIC
.PRINT TRAN V(3) I(VTST0)

** AC SMALL SIGNAL RESONANCE
.AC LIN {TOT} {FLLIM} {FHLIM}

** AC SMALL SIGNAL RESONANCE
.PRINT AC V(4)

** AC SMALL SIGNAL S PARAMETER
** S11 S22
.PRINT AC V(2)
** S21 S12
.PRINT AC V(3)


.END
```

The values of RMS current through and RMS voltage across the planar spiral inductor (on lossy substrate) are obtained with SPICE [2–5] transient analysis. As in the previous design examples, a separate SPICE [2–5] netlist was generated for each selected frequency in the band at 100 MHz–3 GHz. The raw transient analysis output was processed with the supplied C computer language [1] executable *rmscalc*, to compute the RMS current and voltage values for each frequency value in the selected range. These RMS values, as well as the frequency-dependent series resistance for that frequency, are listed in tabular format below:

.

```
1.0E+8,7.063749e+00,4.194293e-01,4.520617
2.0E+8,7.063806e+00,2.099045e-01,8.932182
3.0E+8,7.063826e+00,1.399778e-01,13.343944
4.0E+8,7.063834e+00,1.049986e-01,17.755809
5.0E+8,7.063841e+00,8.400589e-02,22.16774
6.0E+8,7.063844e+00,7.000871e-02,26.579718
7.0E+8,7.063847e+00,6.000965e-02,30.991731
8.0E+8,7.063849e+00,5.250977e-02,35.773403
9.0E+8,7.063850e+00,4.667617e-02,39.815838
1.0E+9,7.063852e+00,4.200908e-02,44.227922
1.1E+9,7.063852e+00,3.819038e-02,48.640023
1.2E+9,7.063854e+00,3.500804e-02,53.052138
1.3E+9,7.063854e+00,3.231520e-02,57.464266
1.4E+9,7.063855e+00,3.000700e-02,61.876405
1.5E+9,7.063856e+00,2.800651e-02,66.288553
1.6E+9,7.063855e+00,2.625605e-02,70.700711
1.7E+9,7.063857e+00,2.471464e-02,75.112877
1.8E+9,7.063857e+00,2.334151e-02,79.525051
1.9E+9,7.063858e+00,2.211290e-02,83.937231
2.0E+9,7.067458e+00,2.101784e-02,88.349417
2.1E+9,7.063858e+00,2.000666e-02,92.761610
2.2E+9,7.063858e+00,1.909712e-02,97.173808
2.3E+9,7.063859e+00,1.826666e-02,101.586010
```

```
2.4E+9,7.063859e+00,1.750541e-02,105.998218
2.5E+9,7.063859e+00,1.680504e-02,110.419430
2.6E+9,7.063860e+00,1.615854e-02,114.822646
2.7E+9,7.063860e+00,1.555993e-02,119.234866
```

It is clear from calculated values for the RMS current through and voltage across the planar spiral inductor (on a lossy substrate) that the bare bones Greenhouse, enhanced Greenhouse, and the Mohan Wheeler scheme generate consistent results, i.e., **as the operating frequency increases, the frequency-dependent current through the planar spiral inductor decreases, and the frequency-dependent series resistance increases, resulting in the Q factor remaining flat|varying slightly over a huge frequency range.** Figures 3.29, 3.30 and 3.31 show frequency-dependent current, resistance, and Q factor, respectively.

3.8   Hexagonal Planar Spiral Inductor on Lossy Substrate Mohan Wheeler Current Sheet Approach

The supplied C computer language [1] executable *planarindmw* can also generate the SPICE [2–5] netlist for a hexagonal planar spiral inductor on a lossy substrate using the Mohan Wheeler current sheet scheme. Typing

./planarindmw at the Linux|MinGW shell command prompt generates the following help information:

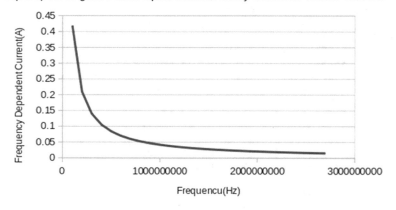

Fig. 3.29 Frequency-dependent current through square|rectangular planar spiral inductor on lossy substrate Mohan Wheeler current sheet scheme

Frequency Dependent Resistance vs. Frequency

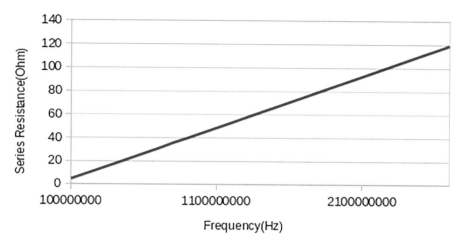

Square|Rectangular Planar Spiral Inductor Lossy Substrate Mohan Wheeler

Fig. 3.30 Frequency-dependent resistance square|rectangular planar spiral inductor on lossy substrate Mohan Wheeler current sheet approach

Q Factor vs. Frequency

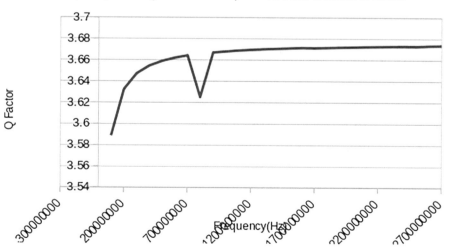

Square|Rectangular Planar Spiral Inductor Mohan Wheeler

Fig. 3.31 Q factor square|rectangular planar spiral inductor lossy substrate Mohan Wheeler current sheet approach

```
./planarindmw
incorrect|insufficient parameters
interactive mode
./planarindmw i|I 1|2|3
batch/command line argument mode
./planarindmw b|B|c|C 1|2|3
<inner diameter(mm)>
<outer diameter(mm)>
<operating frequency(MHz)>
<target inductance value(nH)>
<trace separation(mm)>
<substrate height(mm)>
<oxide height(mm)>
<magnetic material height(mm)>
<buried trace depth(mm)>
sample batch/command line input planar square|rectangle inductor
./planarindmw b 1 20 45 350 25 0.5 2 0.5 1 0.25
sample batch/command line input planar hexagon inductor
./planarindmw C 2 15 35 500 25 0.5 1.75 0.5 1 0.25
1 -> square|rectangle planar spiral
2 -> hexagon planar spiral
3 -> octagon planar spiral
```

Using the supplied sample command line argument input with an operating
frequency of 100 MHz generates the following output:

```
./planarindmw C 2 15 35 100 25 0.5 1.75 0.5 1 0.25
hexagon planar spiral inductor
computed physical parameters
number of turns Wheeler 0.932166
number of turns Mohan 0.914846
skin depth 1.149126e-05 m
trace thickness 1.723689e-05 m
trace width 2.872815e-05 m
average conductor length 1.300298e-01 m
final trace separation 5.000000e-04 m
computed small signal model components
substrate capacitance 1.240861e-10 F
series capacitance 0.000000e+00 F
underpass capacitance 1 1.225254e-16 F
underpass capacitance 2 2.255022e-15 F
oxide capacitance 1.487333e-16 F
series resistance 4.464009 Ohm
substrate resistance 599649.457450 Ohm
magnetic layer resistance 0.000005 Ohm
buried trace resistance 0.001167 Ohm
SPICE netlist planarindmw.cir
```

The contents of the text SPICE [2–5] netlist are listed next:

```
SPIRAL INDUCTOR WHEELER MOHAN

.PARAMS L=2.500000e-08 CSUB=6.204303e-11 CSER=0.000000e+00
.PARAMS CAPU1=1.225254e-16 CAPU2=2.255022e-15 COX=1.487333e-16
.PARAMS RSER=4.464009e+00 RSUB=2.998247e+05 RMAG=3.212408e-06
RBURCOND=0.001167
.PARAMS FREQ=1.000000e+08 FLLIM='0.01*FREQ'
+ FHLIM='1.75*FREQ' CT=1.0e-15 AMPL=10
+ R=50.0 RSMALL=0.001 TS='1.0/(20.0*FREQ)'
+ TSTOP='50.0/FREQ' TSTRT='1.0/FREQ' TOT=100000
.PARAMS RSS=50.0 RSL=50.0 CL=10.0

.SUBCKT PLANARIND2 1 2
** 1 IN
** 2 OUT
CSER 1 3 {CSER}
COX1 1 6 {COX}
COX2 3 4 {COX}
CSUB1 7 0 {CSUB}
CSUB2 5 0 {CSUB}
CU1 3 2 {CAPU1}
CU2 2 4 {CAPU2}
L0 1 8 {L}
RSER 8 3 {RSER}
RMAG1 6 7 {RMAG}
RMAG2 4 5 {RMAG}
RSUB1 7 0 {RSUB}
RSUB2 5 0 {RSUB}
RBC 3 2 {RBURCOND}
.ENDS

** COMMENT OUT APPROPRIATE SECTIONS TO EXECUTE

** TRANSIENT ANALYSIS Q FACTOR
R0 1 2 {RSMALL}
VTST0 2 3 DC 0.0 AC 0.0
XSIMW 3 0 PLANARIND2
VSIG 1 0 DC 0.001
+ SIN(0 {AMPL} {FREQ} 0 0 0)

** AC SMALL SIGNAL ANALYSIS RESONANCE
RS 1 2 {R}
CT 3 4 {CT}
RL 4 0 {R}
XL0 2 3 PLANARIND2
VSIG 1 0 DC 0.001 AC {AMPL}

** AC SMALL SIGNAL ANALYSIS S PARAMETER
RS 1 2 {R}
RL 3 0 {R}
** S11 S21
```

```
XL0 2 3 PLANARIND2
** S22 S12
XL0 3 2 PLANARIND2
VSIG 1 0 DC 0.001 AC 1.0

.OPTIONS METHOD=GEAR NOPAGE RELTOL=1m MINBREAK=4ps

** TRANSIENT ANALYSIS
.IC
.TRAN {TS} {TSTOP} {TSTRT} UIC
.PRINT TRAN V(3) I(VTST0)
.PRINT TRAN V(3) I(VTST0)

** AC SMALL SIGNAL RESONANCE
.PRINT AC V(4)

** AC SMALL SIGNAL S PARAMETER
** S11 S22
.PRINT AC V(2)
** S21 S12
.PRINT AC V(3)

.END
```

SPICE [2–5] transient analysis is used to determine the RMS current through and voltage across the hexagonal planar spiral inductor on lossy substrate, for each selected frequency in the range 100.0 MHz–2.0 GHz. The raw SPICE [2–5] transient analysis output for each selected frequency is processed with the supplied C computer language [1] executable *rmscalc*, and the final results are listed in tabular format below. The last number in each row is the frequency-dependent series resistance at that frequency:

```
1.0E+8,7.063748e+00,4.198173e-01,4.464009
2.0E+8,7.063806e+00,2.100935e-01,8.821814
3.0E+8,7.063825e+00,1.401026e-01,13.178499
4.0E+8,7.063836e+00,1.050917e-01,17.536071
5.0E+8,7.063841e+00,8.408027e-02,21.893749
6.0E+8,7.063845e+00,7.007052e-02,26.251503
7.0E+8,7.063848e+00,6.006254e-02,30.609315
8.0E+8,7.063849e+00,5.255599e-02,34.967173
9.0E+8,7.063851e+00,4.671723e-02,39.325069
1.0E+9,7.063852e+00,4.204599e-02,43.682995
1.1E+9,7.063854e+00,3.822393e-02,48.040949
1.2E+9,7.063854e+00,3.503877e-02,52.398925
1.3E+9,7.063854e+00,3.234357e-02,56.756922
1.4E+9,7.063855e+00,3.003333e-02,61.114937
1.5E+9,7.063856e+00,2.803108e-02,65.472968
1.6E+9,7.063857e+00,2.627908e-02,69.831014
1.7E+9,7.063857e+00,2.473638e-02,74.189073
1.8E+9,7.063857e+00,2.336204e-02,78.547144
1.9E+9,7.063857e+00,2.213234e-02,82.905226
```

```
2.0E+9,7.063857e+00,2.102559e-02,87.263318
2.1E+9,7.063858e+00,2.002425e-02,91.621420
2.2E+9,7.063858e+00,1.911391e-02,95.979530
2.3E+9,7.063858e+00,1.828273e-02,100.337649
2.4E+9,7.063858e+00,1.752080e-02,104.695775
2.5E+9,7.063859e+00,1.681982e-02,109.053909
2.6E+9,7.063859e+00,1.617275e-02,113.412049
2.7E+9,7.063859e+00,1.557361e-02,117.770196
2.8E+9,7.063859e+00,1.501725e-02,122.128349
2.9E+9,7.063859e+00,1.449925e-02,126.486507
```

The hexagonal planar spiral inductor on a lossy substrate, based on the Mohan Wheeler scheme, has very similar performance characteristics as that of the square/ rectangular planar spiral inductor on a lossy substrate, using the same Mohan Wheeler approach. The frequency-dependent current, resistance, and Q factor are shown in Figs. 3.32, 3.33 and 3.34. These have been obtained from the transient analysis data that is listed in tabular format above.

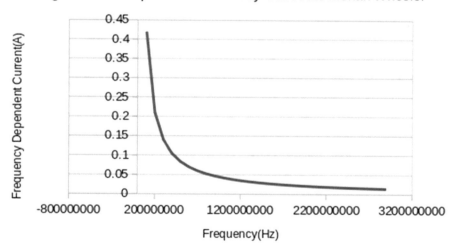

Fig. 3.32 Frequency-dependent current hexagonal planar spiral inductor on lossy substrate Mohan Wheeler scheme

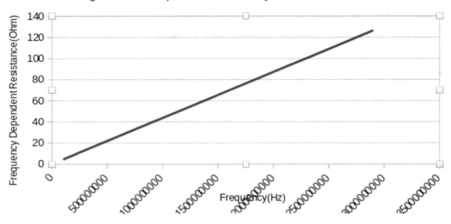

Fig. 3.33 Frequency-dependent resistance hexagonal planar spiral inductor on lossy substrate Mohan wheeler scheme

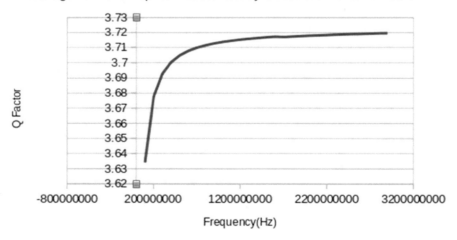

Fig. 3.34 Q factor hexagonal planar spiral inductor on lossy substrate Mohan Wheeler scheme

3.9   Balun-Type Square|Rectangular Planar Spiral Inductor Transformer

The supplied C computer language [1] executable *planarindxfrmr* accepts a predefined set of parameters and generates the values for the components of the equivalent electrical circuit for a planar spiral inductor balun-type integrated transformer, and formats the results as a text SPICE [2–5] netlist. Physically, the transformer consists of four coupled pairs of planar rectangular inductors, arranged as a spiral, in the balun configuration. *The substrate loss is minimum.* Typing ./ **planarindxfrmr** at the Linux|MinGW shell command prompt generates the following help information:

```
./planarindxfrmr
incorrect|insufficient arguments
interactive mode
./planarindxfrmr i|I
<batch/command line argument mode
./planarindxfrmr b|B|c|C
<operating frequency(MHz)>
<maximum square spiral length(mm)>
<maximum square spiral width(mm)>
<trace width(mm)>
<vertical trace separation(mm)>
<substrate thickness(mm)>
<trace thickness(mm)>
<relative permeability of trace conductor>
<substrate relative permittivity>
sample batch|command line input - 400 MHz transformer
./planarindxfrmr c 400 100 80 0.5 1 3 0.001 9 8
```

Using the supplied command line argument input with an operating frequency of 100 MHz generates the following output:

```
./planarindxfrmr c 100 100 80 0.5 1 3 0.001 9 8
total frequency dependent series resistance 12.95571 Ohm
SPICE netlist planarindxfrmr.cir
```

The contents of the generated SPICE [2–5] text netlist are as below:

```
PLANAR TRANSFORMER

.PARAMS CT=1.0E-15 FREQ=1.000000e+08 FLLIM='0.01*FREQ'
FHLIM='1.75*FREQ'
+ RSMALL=0.001 TS='1.0/(15.0*FREQ)'
+ TSTOP='50.0/FREQ' TSTRT='1.0/FREQ' TOT=10000
+ AMPL=10.0 RS=50.0 RB=1.0E+1

.SUBCKT PLTXFRMRBALUNUNIT0 1 2 3 4
** 1 IN 1
```

```
** 2 IN 2
** 3 OUT 1
** 4 OUT 2
C0 3 0 5.902667e-16
C1 4 0 5.902667e-16
C2 3 4 3.541600e-15
L0 1 5 5.016701e-07
L1 2 6 5.016701e-07
k0 L0 L1 0.99
R0 5 3 3.623974e+00
R1 6 4 3.623974e+00
R2 1 0 2.000000e+20
R3 2 0 2.000000e+20
.ENDS


.SUBCKT PLTXFRMRBALUNUNIT1 1 2 3 4
** 1 IN 1
** 2 IN 2
** 3 OUT 1
** 4 OUT 2
C0 3 0 4.692620e-16
C1 4 0 4.692620e-16
C2 3 4 2.815572e-15
L0 1 5 5.021007e-07
L1 2 6 5.021007e-07
k0 L0 L1 0.99
R0 5 3 2.881059e+00
R1 6 4 2.881059e+00
R2 1 0 1.590000e+20
R3 2 0 1.590000e+20
.ENDS


.SUBCKT PLTXFRMRBALUNUNIT2 1 2 3 4
** 1 IN 1
** 2 IN 2
** 3 OUT 1
** 4 OUT 2
C0 3 0 5.843640e-16
C1 4 0 5.843640e-16
C2 3 4 3.506184e-15
L0 1 5 5.016869e-07
L1 2 6 5.016869e-07
k0 L0 L1 0.99
R0 5 3 3.587734e+00
R1 6 4 3.587734e+00
R2 1 0 1.980000e+20
R3 2 0 1.980000e+20
.ENDS


.SUBCKT PLTXFRMRBALUNUNIT3 1 2 3 4
** 1 IN 1
** 2 IN 2
** 3 OUT 1
```

```
** 4 OUT 2
C0 3 0 4.663107e-16
C1 4 0 4.663107e-16
C2 3 4 2.797864e-15
L0 1 5 5.021140e-07
L1 2 6 5.021140e-07
k0 L0 L1 0.99
R0 5 3 2.862939e+00
R1 6 4 2.862939e+00
R2 1 0 1.580000e+20
R3 2 0 1.580000e+20
.ENDS

.SUBCKT PLTXFRMRSTACKUNIT0 1 2 3 4
** 1 IN 1
** 2 IN 2
** 3 OUT 1
** 4 OUT 2
C1 4 0 5.902667e-16
C2 3 4 3.541600e-15
L0 1 5 5.016701e-07
L1 2 6 5.016701e-07
k0 L0 L1 0.99
R0 5 3 3.623974e+00
R1 6 4 3.623974e+00
R3 2 0 2.000000e+20
.ENDS

.SUBCKT PLTXFRMRSTACKUNIT1 1 2 3 4
** 1 IN 1
** 2 IN 2
** 3 OUT 1
** 4 OUT 2
C1 4 0 4.692620e-16
C2 3 4 2.815572e-15
L0 1 5 5.021007e-07
L1 2 6 5.021007e-07
k0 L0 L1 0.99
R0 5 3 2.881059e+00
R1 6 4 2.881059e+00
R3 2 0 1.590000e+20
.ENDS

.SUBCKT PLTXFRMRSTACKUNIT2 1 2 3 4
** 1 IN 1
** 2 IN 2
** 3 OUT 1
** 4 OUT 2
C1 4 0 5.843640e-16
C2 3 4 3.506184e-15
L0 1 5 5.016869e-07
L1 2 6 5.016869e-07
k0 L0 L1 0.99
```

```
R0 5 3 3.587734e+00
R1 6 4 3.587734e+00
R3 2 0 1.980000e+20
.ENDS


.SUBCKT PLTXFRMRSTACKUNIT3 1 2 3 4
** 1 IN 1
** 2 IN 2
** 3 OUT 1
** 4 OUT 2
C1 4 0 4.663107e-16
C2 3 4 2.797864e-15
L0 1 5 5.021140e-07
L1 2 6 5.021140e-07
k0 L0 L1 0.99
R0 5 3 2.862939e+00
R1 6 4 2.862939e+00
R3 2 0 1.580000e+20
.ENDS


.SUBCKT BALUNTRANSFORMER 1 2 3 4
** 1 IN 1
** 2 IN 2
** 3 OUT 1
** 4 OUT 2
XBTU0 1 2 5 6 PLTXFRMRBALUNUNIT0
XBTU1 5 6 7 8 PLTXFRMRBALUNUNIT1
XBTU2 7 8 9 10 PLTXFRMRBALUNUNIT2
XBTU3 9 10 3 4 PLTXFRMRBALUNUNIT3
.ENDS


.SUBCKT STACKTRANSFORMER 1 2 3 4
** 1 IN 1
** 2 IN 2
** 3 OUT 1
** 4 OUT 2
XSTU0 1 2 5 6 PLTXFRMRSTACKUNIT0
XSTU1 5 6 7 8 PLTXFRMRSTACKUNIT1
XSTU2 7 8 9 10 PLTXFRMRSTACKUNIT2
XSTU3 9 10 3 4 PLTXFRMRSTACKUNIT3
.ENDS


** COMMENT OUT BALUN TYPE, STACK TYPE As NEEDED
** TRANSIENT ANALYSIS
R0 1 2 {RS}
R1 4 0 {RS}
R2 3 0 {RB}
R3 5 0 {RB}
XBTXRFMR 2 0 4 0 BALUNTRANSFORMER

XSTXRFMR 2 0 4 0 STACKTRANSFORMER
```

```
VSIG 1 0 DC 0.01 SIN(0 {AMPL} {FREQ} 0 0 0)


** TRANSIENT ANALYSIS Q FACTOR
R0 1 2 {RSMALL}
VTST0 2 3 DC 0.0 AC 0.0
XBTXRFMR 3 0 0 0 BALUNTRANSFORMER

XSTXRFMR 3 0 0 0 STACKTRANSFORMER

VSIG 1 0 DC 0.01 SIN(0 {AMPL} {FREQ} 0 0 0)

** AC SMALL SIGNAL SELF RESONANCE
RS 1 2 {RS}
RL 3 0 {RS}
** BALUN TYPE
XBTXRFMR 2 0 3 0 BALUNTRANSFORMER
** STACK TYPE
XSTXRFMR 2 0 3 0 STACKTRANSFORMER
VSIG 1 0 DC 0.001 AC {AMPL}

** AC SMALL SIGNAL S PARAMETER
RS 1 2 {RS}
RL 3 0 {RS}
** BALUN TYPE S11 S21
XBTXRFMR 2 0 3 0 BALUNTRANSFORMER
** BALUN TYPE S22 S12
XBTXRFMR 3 0 2 0 BALUNTRANSFORMER
** STACK TYPE S11 S21
XSTXRFMR 2 0 3 0 STACKTRANSFORMER
** STACK TYPE S22 S12
XSTXRFMR 3 0 2 0 STACKTRANSFORMER
VSIG 1 0 DC 0.001 AC 1.0

.OPTIONS METHOD=GEAR NOPAGE RELTOL=1m MINBREAK=5ps
.IC
.TRAN {TS} {TSTOP} {TSTRT} UIC
** TRANSIENT ANALYSIS
.PRINT TRAN V(2) V(10)
** TRANSIENT ANALYSIS Q FACTOR
.PRINT TRAN V(3) I(VTST0)

.AC LIN {TOT} {FLLIM} {FHLIM}
** AC SMALL SIGNAL SELF RESONANCE
.PRINT AC V(3)

** AC SMALL SIGNAL S PARAMETER
** S11 S22
.PRINT AC V(2)
** S21 S12
.PRINT AC V(3)

.END
```

SPICE [2–5] transient analysis is used to determine the RMS current through and voltage drop across the planar spiral inductor balun-type transformer in the frequency range of 100 MHz–3 GHz. As the frequency-dependent series resistance of each planar spiral inductor changes, a separate SPICE [2–5] netlist is generated for each selected frequency in the frequency band. The raw SPICE [2–5] transient analysis output for each selected frequency is processed using the supplied C computer language [1] executable **rmscalc** that calculates the RMS current and voltage for that frequency. These are listed below in tabular format. As in all the previous design examples, the number in the first column represents the frequency, the second number the RMS voltage, the third the R < S current, and the last number the frequency-dependent series resistance:

```
1.0E+8,7.064295e+00,1.894819e-01,12.95571
2.0E+8,7.065663e+00,1.184045e-01,13.36034
3.0E+8,7.061438e+00,8.371200e-02,13.67628
4.0E+8,7.061449e+00,6.424473e-02,13.94628
5.0E+8,7.061455e+00,5.200435e-02,14.18693
6.0E+8,7.061459e+00,4.362271e-02,14.40672
7.0E+8,7.066255e+00,3.752439e-02,14.61071
8.0E+8,7.061461e+00,3.292971e-02,14.80217
9.0E+8,7.061463e+00,2.932580e-02,14.98340
1.0E+9,7.061464e+00,2.642632e-02,14.15606
1.1E+9,7.061566e+00,2.402681e-02,15.32140
1.2E+9,7.066260e+00,2.203205e-02,15.49040
1.3E+9,7.061482e+00,2.033158e-02,15.63383
1.4E+9,7.061809e+00,1.886748e-02,15.78233
1.5E+9,7.058496e+00,1.759749e-02,15.92642
1.6E+9,7.059728e+00,1.649035e-02,16.06652
1.7E+9,7.061102e+00,1.553798e-02.16.203
1.8E+9,7.051044e+00,1.470538e-02,16.43318
1.9E+9,7.047905e+00,1.391154e-02,16.46603
2.0E+9,7.054484e+00,1.318566e-02,16.59367
2.1E+9,7.057736e+00,1.256627e-02,16.71843
2.2E+9,7.048010e+00,1.199471e-02,16.84078
2.3E+9,7.043871e+00,1.145387e-02,16.96087
2.4E+9,7.044361e+00,1.096731e-02,17.0878
2.5E+9,7.052149e+00,1.048404e-02,17.19487
2.6E+9,7.045855e+00,1.006628e-02,17.30995
2.7E+9,7.033381e+00,9.763193e-03,17.42541
2.8E+9,7.024356e+00,9.369398e-03,17.53232
2.9E+9,7.038242e+00,9.017792e-03,17.64137
3.0E+9,7.023954e+00,8.711776e-03,17.74906
```

The frequency-dependent current through the transformer, the frequency-dependent resistance, and the Q factor are shown in Figs. 3.35, 3.36 and 3.37, respectively.

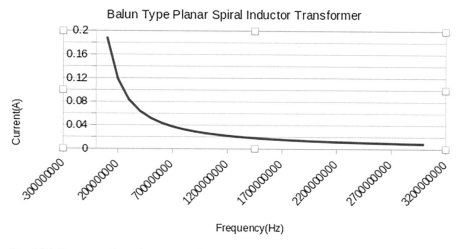

Fig. 3.35 Frequency-dependent current planar spiral inductor balun-type transformer

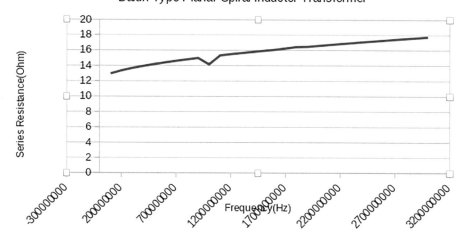

Fig. 3.36 Frequency-dependent resistance planar spiral inductor balun-type transformers

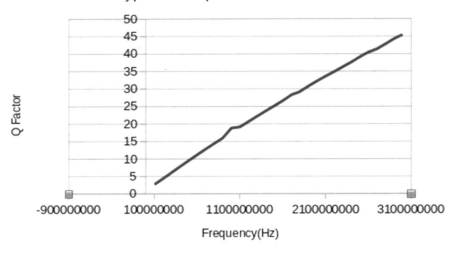

Fig. 3.37 Q factor planar spiral inductor balun-type transformers

3.10   Stack-Type Square|Rectangular Planar Spiral Inductor

Transformer

The supplied C computer language [1] executable ***planarindxfrmr*** accepts a predefined set of parameters and generates the values for the components of the equivalent electrical circuit for a planar spiral inductor stack-type transformer, and formats the results as a text SPICE [2–5] netlist. Physically, the transformer consists of four coupled pairs of planar rectangular inductors, arranged as a stacked spiral. *The substrate is minimum loss.* Typing **./planarindxfrmr** at the Linux|MinGW shell command prompt generates the following help information:

```
./planarindxfrmr
incorrect|insufficient arguments
interactive mode
./planarindxfrmr i|I
<batch/command line argument mode
./planarindxfrmr b|B|c|C
<operating frequency(MHz)>
<maximum square spiral length(mm)>
<maximum square spiral width(mm)>
<trace width(mm)>
<vertical trace separation(mm)>
<substrate thickness(mm)>
```

```
<trace thickness (mm) >
<relative permeability of trace conductor>
<substrate relative permittivity>
sample batch|command line input - 400 MHz transformer
./planarindxfrmr c 400 100 80 0.5 1 3 0.001 9 8
```

Using the supplied command line argument input with an operating frequency of 100 MHz generates the following output:

```
./planarindxfrmr c 100 100 80 0.5 1 3 0.001 9 8
total frequency dependent series resistance 12.95571 Ohm
SPICE netlist planarindxfrmr.cir
```

The contents of the generated SPICE [2–5] text netlist are as below:

```
PLANAR TRANSFORMER

.PARAMS CT=1.0E-15 FREQ=1.000000e+08 FLLIM='0.01*FREQ'
FHLIM='1.75*FREQ'
+ RSMALL=0.001 TS='1.0/(15.0*FREQ)'
+ TSTOP='50.0/FREQ' TSTRT='1.0/FREQ' TOT=10000
+ AMPL=10.0 RS=50.0 RB=1.0E+1

.SUBCKT PLTXFRMRBALUNUNIT0 1 2 3 4
** 1 IN 1
** 2 IN 2
** 3 OUT 1
** 4 OUT 2
C0 3 0 5.902667e-16
C1 4 0 5.902667e-16
C2 3 4 3.541600e-15
L0 1 5 5.016701e-07
L1 2 6 5.016701e-07
k0 L0 L1 0.99
R0 5 3 3.623974e+00
R1 6 4 3.623974e+00
R2 1 0 2.000000e+20
R3 2 0 2.000000e+20
.ENDS

.SUBCKT PLTXFRMRBALUNUNIT1 1 2 3 4
** 1 IN 1
** 2 IN 2
** 3 OUT 1
** 4 OUT 2
C0 3 0 4.692620e-16
C1 4 0 4.692620e-16
C2 3 4 2.815572e-15
L0 1 5 5.021007e-07
L1 2 6 5.021007e-07
k0 L0 L1 0.99
```

```
R0 5 3 2.881059e+00
R1 6 4 2.881059e+00
R2 1 0 1.590000e+20
R3 2 0 1.590000e+20
.ENDS


.SUBCKT PLTXFRMRBALUNUNIT2 1 2 3 4
** 1 IN 1
** 2 IN 2
** 3 OUT 1
** 4 OUT 2
C0 3 0 5.843640e-16
C1 4 0 5.843640e-16
C2 3 4 3.506184e-15
L0 1 5 5.016869e-07
L1 2 6 5.016869e-07
k0 L0 L1 0.99
R0 5 3 3.587734e+00
R1 6 4 3.587734e+00
R2 1 0 1.980000e+20
R3 2 0 1.980000e+20
.ENDS


.SUBCKT PLTXFRMRBALUNUNIT3 1 2 3 4
** 1 IN 1
** 2 IN 2
** 3 OUT 1
** 4 OUT 2
C0 3 0 4.663107e-16
C1 4 0 4.663107e-16
C2 3 4 2.797864e-15
L0 1 5 5.021140e-07
L1 2 6 5.021140e-07
k0 L0 L1 0.99
R0 5 3 2.862939e+00
R1 6 4 2.862939e+00
R2 1 0 1.580000e+20
R3 2 0 1.580000e+20
.ENDS


.SUBCKT PLTXFRMRSTACKUNIT0 1 2 3 4
** 1 IN 1
** 2 IN 2
** 3 OUT 1
** 4 OUT 2
C1 4 0 5.902667e-16
C2 3 4 3.541600e-15
L0 1 5 5.016701e-07
L1 2 6 5.016701e-07
k0 L0 L1 0.99
R0 5 3 3.623974e+00
R1 6 4 3.623974e+00
```

```
R3 2 0 2.000000e+20
.ENDS


.SUBCKT PLTXFRMRSTACKUNIT1 1 2 3 4
** 1 IN 1
** 2 IN 2
** 3 OUT 1
** 4 OUT 2
C1 4 0 4.692620e-16
C2 3 4 2.815572e-15
L0 1 5 5.021007e-07
L1 2 6 5.021007e-07
k0 L0 L1 0.99
R0 5 3 2.881059e+00
R1 6 4 2.881059e+00
R3 2 0 1.590000e+20
.ENDS


.SUBCKT PLTXFRMRSTACKUNIT2 1 2 3 4
** 1 IN 1
** 2 IN 2
** 3 OUT 1
** 4 OUT 2
C1 4 0 5.843640e-16
C2 3 4 3.506184e-15
L0 1 5 5.016869e-07
L1 2 6 5.016869e-07
k0 L0 L1 0.99
R0 5 3 3.587734e+00
R1 6 4 3.587734e+00
R3 2 0 1.980000e+20
.ENDS


.SUBCKT PLTXFRMRSTACKUNIT3 1 2 3 4
** 1 IN 1
** 2 IN 2
** 3 OUT 1
** 4 OUT 2
C1 4 0 4.663107e-16
C2 3 4 2.797864e-15
L0 1 5 5.021140e-07
L1 2 6 5.021140e-07
k0 L0 L1 0.99
R0 5 3 2.862939e+00
R1 6 4 2.862939e+00
R3 2 0 1.580000e+20
.ENDS
```

```
.SUBCKT BALUNTRANSFORMER 1 2 3 4
** 1 IN 1
** 2 IN 2
** 3 OUT 1
** 4 OUT 2
XBTU0 1 2 5 6 PLTXFRMRBALUNUNIT0
XBTU1 5 6 7 8 PLTXFRMRBALUNUNIT1
XBTU2 7 8 9 10 PLTXFRMRBALUNUNIT2
XBTU3 9 10 3 4 PLTXFRMRBALUNUNIT3
.ENDS

.SUBCKT STACKTRANSFORMER 1 2 3 4
** 1 IN 1
** 2 IN 2
** 3 OUT 1
** 4 OUT 2
XSTU0 1 2 5 6 PLTXFRMRSTACKUNIT0
XSTU1 5 6 7 8 PLTXFRMRSTACKUNIT1
XSTU2 7 8 9 10 PLTXFRMRSTACKUNIT2
XSTU3 9 10 3 4 PLTXFRMRSTACKUNIT3
.ENDS

** COMMENT OUT BALUN TYPE, STACK TYPE As NEEDED
** TRANSIENT ANALYSIS
R0 1 2 {RS}
R1 4 0 {RS}
R2 3 0 {RB}
R3 5 0 {RB}
XBTXRFMR 2 0 4 0 BALUNTRANSFORMER

XSTXRFMR 2 0 4 0 STACKTRANSFORMER

VSIG 1 0 DC 0.01 SIN(0 {AMPL} {FREQ} 0 0 0)

** TRANSIENT ANALYSIS Q FACTOR
R0 1 2 {RSMALL}
VTST0 2 3 DC 0.0 AC 0.0
XBTXRFMR 3 0 0 0 BALUNTRANSFORMER

XSTXRFMR 3 0 0 0 STACKTRANSFORMER

VSIG 1 0 DC 0.01 SIN(0 {AMPL} {FREQ} 0 0 0)

** AC SMALL SIGNAL SELF RESONANCE
RS 1 2 {RS}
RL 3 0 {RS}
** BALUN TYPE
XBTXRFMR 2 0 3 0 BALUNTRANSFORMER
** STACK TYPE
XSTXRFMR 2 0 3 0 STACKTRANSFORMER
VSIG 1 0 DC 0.001 AC {AMPL}
```

```
** AC SMALL SIGNAL S PARAMETER
RS 1 2 {RS}
RL 3 0 {RS}
** BALUN TYPE S11 S21
XBTXRFMR 2 0 3 0 BALUNTRANSFORMER
** BALUN TYPE S22 S12
XBTXRFMR 3 0 2 0 BALUNTRANSFORMER
** STACK TYPE S11 S21
XSTXRFMR 2 0 3 0 STACKTRANSFORMER
** STACK TYPE S22 S12
XSTXRFMR 3 0 2 0 STACKTRANSFORMER
VSIG 1 0 DC 0.001 AC 1.0

.OPTIONS METHOD=GEAR NOPAGE RELTOL=1m MINBREAK=5ps
.IC
.TRAN {TS} {TSTOP} {TSTRT} UIC
** TRANSIENT ANALYSIS
.PRINT TRAN V(2) V(10)
** TRANSIENT ANALYSIS Q FACTOR
.PRINT TRAN V(3) I(VTST0)

.AC LIN {TOT} {FLLIM} {FHLIM}
 ** AC SMALL SIGNAL SELF RESONANCE
.PRINT AC V(3)

** AC SMALL SIGNAL S PARAMETER
** S11 S22
.PRINT AC V(2)
** S21 S12
.PRINT AC V(3)

.END
```

The TMS current through and the RMS voltage across the stack-type planar spiral inductor transformer, obtained using the same procedure as in the case of the balun-type planar spiral inductor transformer, are listed below:

```
1.0E+8,7.064214e+00,1.894851e-01,12.95571
2.0E+8,7.063179e+00,1.184259e-01,13.36034
3.0E+8,7.061439e+00,8.372243e-02,13.67628
4.0E+8,7.061450e+00,6.425083e-02,13.92864
5.0E+8,7.061455e+00,5.198784e-02,14.18693
6.0E+8,7.061458e+00,4.360976e-02,14.40672
7.0E+8,7.061460e+00,3.754911e-02,14.61071
8.0E+8,7.061462e+00,3.293733e-02,14.80217
9.0E+8,7.061463e+00,2.933323e-02,14.98318
1.0E+9,7.061464e+00,2.642397e-02,15.15606
1.1E+9,7.066260e+00,2.404773e-02,15.3214
1.2E+8,7.061464e+00,2.205932e-02,15.4804
1.3E+9,7.061465e+00,2.037601e-02,15.63383
1.4E+9,7.061472e+00,1.890802e-02,15.78233
1.5E+9,7.061560e+00,1.765270e-02,15.92642
```

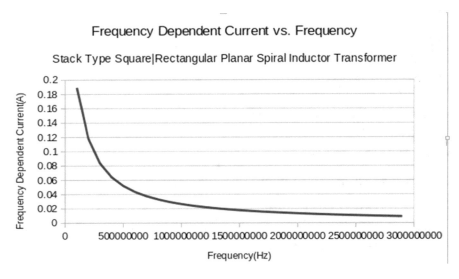

Fig. 3.38 Frequency-dependent current stack-type planar spiral inductor transformers

```
1.6E+9,7.061502e+00,1.655106e-02,16.06652
1.7R+9,7.061466e+00,1.557767e-02,16.203
1.8E+9,7.061480e+00,1.469228e-02,16.33618
1.9E+9,7.061535e+00,1.391141e-02,16.46633
2.0E+9,7.061543e+00,1.320725e-02,16.59367
2.1E+9,7.061465e+00,1.257796e-02,16.71843
2.2E+9,7.061465e+00,1.199442e-02,16.84078
2.3E+9,7.061475e+00,1.145957e-02,16.96087
2.4E+9,7.061711e+00,1.096186e-02,17.07887
2.5E+9,7.061640e+00,1.051229e-02,17.17087
2.6E+9,7.061675e+00,1.008852e-02,17.30905
2.7E+9,7.061777e+00,9.702949e-03,17.42145
2.8E+9,7.061567e+00,9.333881e-03,17.5322
2.9E+9,7.061656e+00,8.994869e-03,17.64137
```

The frequency-dependent current, resistance, and the computed Q factor are shown in Figs. 3.38, 3.39 and 3.40.

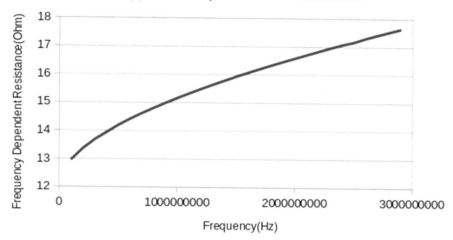

Fig. 3.39 Frequency-dependent resistance planar spiral inductor transformers

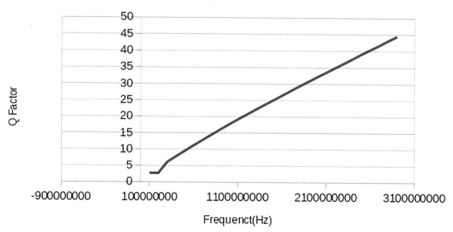

Fig. 3.40 Q factor stack-type planar spiral inductor transformers

3.11   Square|Rectangular Planar Spiral Inductor Transformer on Lossy Substrate

The supplied C computer language [1] executable *planarindgh* generates the SPICE [2–5] text input format netlist for the equivalent electrical circuit for square|rectangular planar spiral inductor transformer on a lossy substrate. Typing *./planarindgh* at the Linux|MinGW shell command prompt generates the following help information:

```
./planarindgh
incorrect/insufficient arguments
interactive mode
./spiralind_1 i|I 1|2|3
batch/command line argument mode
./planarindgh b|B|c|C 1|2|3
<trace width(mm)>
<trace separation(mm)>
<oxide thickness(mm)>
<substrate thickness(mm)>
<trace thickness(mm)>
<target inductance value(nH)>
<operating frequency(MHz)>
<inductor maximum length(mm)>
<inductor maximum width(mm)>
<maximum outer diameter(mm)>
maximum length, width non-zero ONLY
<under pass depth(mm)
<tolerance e.g., 0.95>
tolerance determines how close the
computed inductance(self + mutual)
value is to the specified target value
for square, rectangle spiral inductor
maximum spiral outer diameter non-zero
ONLY for hexagon, octagon spiral inductor
1 -> rectangle|square spiral inductor
2 -> hexagon spiral inductor
3 -> octagon spiral inductor
sample command line input square|rectangle spiral inductor
./planarindgh b 1 1.25 0.5 0.45 1.75 0.001 30 200 60 70 0 0.5 0.95
sample command line input hexagon spiral inductor
./planarindgh C 2 1.25 0.5 0.45 1.75 0.001 100 100 0 0 80 0.5 0.95
sample command line input octagon spiral inductor
./planarindgh B 3 1.25 0.5 0.45 1.75 0.001 150 100 0 0 80 0.5 0.95
Only parallel inductor segment pairs
contribute to mutual inductance
to ensure accurate sequence of input parameters, it is best to execute
the program using command line arguments
```

Using the supplied sample command line argument input generates the following help information:

```
./planarindgh b 1 1.25 0.5 0.45 1.75 0.001 30 200 60 70 0 0.5 0.95
target 3.000000e-08 H   computed maximum 2.756018e-08 H
total number of traces 16
total inductor length 0.808500 mm
skin depth 4.615088e-06 m
frequency dependent series resistance 12.085973 Ohm
substrate resistance 1108.225108 Ohm
series capacitance 1.106750e-13 F
substrate capacitance 1.890461e-08 F
oxide capacitance 1.080263e-11 F
Length(m) self-inductance(H) mutual inductance(H) total inductance(H)
0 6.000000e-02 1.717523e-09 0.000000e+00 1.717523e-09
1 7.000000e-02 2.003777e-09 0.000000e+00 3.721300e-09
2 5.825000e-02 1.667429e-09 0.000000e+00 5.388729e-09
3 6.825000e-02 1.953683e-09 0.000000e+00 7.342412e-09
4 5.475000e-02 1.567240e-09 4.565003e-10 9.366152e-09
5 5.475000e-02 1.567240e-09 4.565003e-10 1.138989e-08
6 5.125000e-02 1.467051e-09 4.204431e-10 1.327739e-08
7 5.125000e-02 1.467051e-09 4.204431e-10 1.516488e-08
8 4.775000e-02 1.366862e-09 3.848645e-10 1.691661e-08
9 4.775000e-02 1.366862e-09 3.848645e-10 1.866833e-08
10 4.425000e-02 1.266673e-09 3.497999e-10 2.028481e-08
11 4.425000e-02 1.266673e-09 3.497999e-10 2.190128e-08
12 4.075000e-02 1.166484e-09 3.152899e-10 2.338305e-08
13 4.075000e-02 1.166484e-09 3.152899e-10 2.486483e-08
14 3.725000e-02 1.066296e-09 2.813824e-10 2.621251e-08
15 3.725000e-02 1.066296e-09 2.813824e-10 2.756018e-08
SPICE netlist planarindgh.cir
```

The contents of the text SPICE [2–5] netlist are:

```
SQUARE|RECTANGULAR PLANAR INDUCTOR LUMPED MODEL

.PARAMS FREQ=2.000000e+08 FLLIM='0.01*FREQ' FHLIM='1.75*FREQ'
+ RSMALL=0.001 TS='1.0/(20.0*FREQ)'
+ TSTOP'50.0/FREQ' TSTRT='1.0/FREQ' TOT=5000
.PARAMS COX=1.080263e-11 CSER=1.106750e-13
.PARAMS CSUB=9.452305e-09 LS=2.756018e-08
.PARAMS RSER=12.085973 RSUB=554.112554 R=50.0 C=1.0E-8 AMPL=10
.PARAMS RSS=50.0 RSL=50.0
.PARAMS CL=10.0 CC=1.0E-12 CTEST=1.0E-15
.SUBCKT PLANARSPIRALIND 1 2
** 1 IN
** 2 OUT
CSER 1 2 {CSER}
COX0 1 4 {COX}
COX1 2 5 {COX}
CSUB0 4 0 {CSUB}
CSUB1 5 0 {CSUB}
```

```
L 1 3 {LS}
RSER 3 2 {RSER}
RSUB0 4 0 {RSUB}
RSUB1 5 0 {RSUB}
.ENDS

.SUBCKT PLANARINDXFRMR 1 2 3 4
** 1 IN 1
** 2 OUT 1
** 3 IN 2
** 4 OUT 2
COX11 1 9 {COX}
COX12 2 10 {COX}
COX21 3 7 {COX}
COX22 4 8 {COX}
CS111 9 0 {CSUB}
CSI12 10 0 {CSUB}
CS121 7 0 {CSUB}
CSI32 8 0 {CSUB}
CSER1 1 2 {CSER}
CSER2 3 4 {CSER}
CCOUP 5 6 {CC}
L1 1 5 {LS}
L2 3 6 {LS}
RSER1 2 5 {RSER}
RSER2 4 6 {RSER}
RSI11 9 0 {RSUB}
RSI12 10 0 {RSUB}
RSI21 7 0 {RSUB}
RSI22 8 0 {RSUB}
k0 L1 L2 0.99
.ENDS

** COMMENT OUT AC OR TRANSIENT ANALYSIS AS NEEDED
** TRANSIENT ANALYSIS Q FACTOR
R0 1 2 {RSMALL}
VTST0 2 3 DC 00 AC 0.0
XPLI 3 0 PLANARSPIRALIND
VSIG 1 0 DC 0.001
+ SIN(0 {AMPL} {FREQ} 0 0 0)

** TRANSIENT ANALYSIS PLANAR SPIRAL INDUCTOR TRANSFORMER
R0 1 2 {RSMALL}
VTST0 2 3 DC 00 AC 0.0
XPLIXFRMR 3 0 0 0 PLANARINDXFRMR
VSIG 1 0 DC 0.001
+ SIN(0 {AMPL} {FREQ} 0 0 0)
```

```
** AC SMALL SIGNAL RESONANCE
CT 2 3 {CTEST}
RS 1 2 {R}
RL 4 0 {R}
XPLI 3 4 PLANARSPIRALIND
VSIG 1 0 DC 0.001 AC {AMPL}

** AC SMALL SIGNAL S PARAMETER
RS 1 2 {R}
RL 3 0 {R}
** S11 S21
XPLI 2 3 PLANARSPIRALIND
** S22 S12
XPLI 3 2 PLANARSPIRALIND
VSIG 1 0 DC 0.001 AC 1.0

.OPTIONS METHOD=GEAR NOPAGE RELTOL=1m
** TRANSIENT ANALYSIS Q FACTOR SPIRAL INDUCTOR AND TRANSFORMER
.IC
.TRAN {TS} {TSTOP} {TSTRT} UIC
.PRINT TRAN V(3) I(VTST0)

.AC LIN {TOT} {FLLIM} {FHLIM}
** AC SMALL SIGNAL RESONANCE
.PRINT AC V(4)

** S PARAMETER S11 S22
.PRINT AC V(2)
** S PARAMETER S21 S12
.PRINT AC V(3)

.END
```

As in all previous design examples, the RMS current through and RMS voltage across the lossy substrate square|rectangular planar spiral inductor transformer are computed by using a separate SPICE [2–5] netlist for each frequency selected from the frequency band of 100.0 MHz–2.0 GHz. A separate SPICE [2–5] netlist is used as the series resistance of each spiral inductor varies with frequency. For each selected frequency, the corresponding RMS current and voltage values are calculated using the supplied C computer language [1] executable **rmscalc**. These RMS currents, voltages, frequency-dependent resistances, and corresponding frequencies are listed in tabular format below:

```
1.0E+8,7.063532e+00,3.342305e-01,11.719932
2.0E+8,7.063562e+00,3.063908e-01,12.085973
3.0E+8,7.063571e+00,3.145938e-01,12.371771
4.0E+8,7.067177e+00,3.360568e-01,12.616022
5.0E+8,7.063584e+00,3.651467e-01,12.833717
6.0E+8,7.063586e+00,3.998910e-01,13.032545
7.0E+8,7.063589e+00,4.385641e-01,13.217071
```

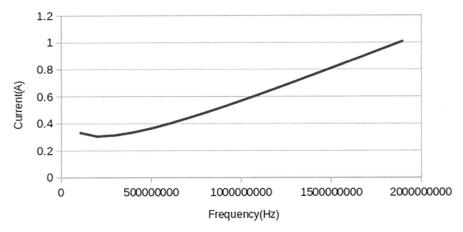

Fig. 3.41 Frequency-dependent current square|rectangular planar spiral inductor transformer lossy substrate

```
8.0E+8,7.063591e+00,4.800941e-01,13.390272
9.0E+8,7.063593e+00,5.237603e-01,13.554213
1.0E+9,7.063594e+00,5.690270e-01,13.710401
1.1E+9,7.063595e+00,6.155324e-01,13.859970
1.2E+9,7.063596e+00,6.630190e-01,14.003894
1.3E+9,7.063596e+00,7.112565e-01,14.104612
1.4E+9,7.063596e+00,7.601081e-01,14.276940
1.5E+9,7.063595e+00,8.094055e-01,14.407208
1.6E+9,7.063594e+00,8.591539e-01,14.534017
1.7E+9,7.063594e+00,9.092350e-01,14.657582
1.8E+9,7.063592e+00,9.595910e-01,14.777958
1.9E+9,7.063590e+00,1.010173e+00,14.895691
```

The frequency-dependent current, resistance, and Q factor are shown in Figs. 3.41, 3.42 and 3.43, respectively. Clearly, the effects of the lossy substrate become prominent as the input signal frequency increases, and both the series resistance (frequency dependent) and the current through the transformer increase.

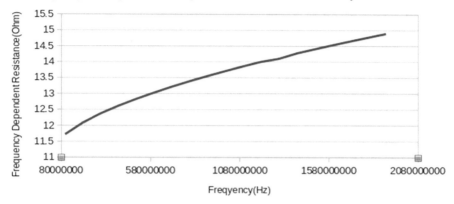

Fig. 3.42 Frequency-dependent resistance square|rectangular planar spiral inductor transformer lossy substrate

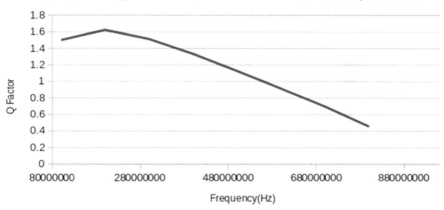

Fig. 3.43 Q factor square|rectangular planar spiral inductor transformer lossy substrate

3.12 Printed Planar Dipole Antenna

The supplied C computer language [1] executable *planarantdp* generates the SPICE [2–5] netlist for a printed (etched on a printed circuit board) planar dipole antenna. All antennas have to be exposed to air, as it needs to transmit|receive electromagnetic waves efficiently. So unlike planar spiral inductors, antennas cannot be embedded inside an integrated circuit. These antennas are called "printed" as they are fabricated on a printed circuit board. Typing **./planarantdp** at the Linux|MinGW shell command prompt generates the following help information:

```
./planarantdp
incorrect|insufficient input
./planarantdp b|B|c|C
<characteristic impedance Ohm>
<dielectric constant - substrate>
<frequency MHz>
<substrate thickness mm>
<trace thickness mm>
sample batch|command line input
./planarantdp B 50.0 9.8 500 1 0.01
```

Using the supplied command line argument input with a frequency of 100 MHz generates the following output:

```
./planarantdp B 50.0 9.8 100 1 0.01
dielectric constant 9.8000
effective dielectric constant 1.0337
antenna half length 0.737676 m
antenna trace width 0.645497 m
antenna half inductor 1.422775e-09 H
antenna half length inductor parasitic capacitor 4.392823e-09 F
antenna capacitor 1.782151e-09 F
antenna half length inductor parasitic resistor 0.000038 Ohm
radiation resistance 12.325 Ohm
approximate trace separation 0.012000 m
approximate antenna capacitor by iteration 4.587197e-10 F
SPICE netlist planarantdp.cir
```

The contents of the text SPICE [2–5] input format netlist are listed below.

```
BALUN|COAX INPUT MATCHED MICROSTRiP DIPOLE ANTENNA - SERIES RLC

.PARAMS C=1.782151e-09 L=1.422775e-09 CSP=4.392823e-09
+ FREQ=1.000000e+08 LT=1.988132e-08 FLLIM='0.1*FREQ'
+ FHLIM='1.75*FREQ'
.PARAMS Z0=50.0000 RP=3.839821e-05 RCOMP=12.325 AMPL=5 TD=1.0E-9
+ TOT=50000 TS='1.0/(15.0*FREQ)' TSTOP='50.0/FREQ' TSTRT='1.0/FREQ'
+ CCOUP=4.587197e-10

** ADJUST|EDIT CONNECTIONS|PARAMETERS TO OPTIMIZE PERFORMANCE

** SUB-CIRCUIT FOR S PARAMETERS
.SUBCKT PLNRDPA 1 2
C0 1 2 {C}
C1 1 2 {CCOUP}
CSP0 1 0 {CSP}
```

```
CSP1 2 0 {CSP}
L0 3 5 {L}
L1 2 6 {L}
R0 5 1 {RP}
R1 6 4 {RP}
RCOMP0 3 0 {RCOMP}
RCOMP1 4 0 {RCOMP}
k0 L0 L1 0.9
.ENDS


** BALUN
.SUBCKT BALUN 1 2 3
** 1 IN
** 2 OUT 1
** 3 OUT 2
LT0 1 0 {LT}
LT1 2 0 {LT}
LT2 0 3 {LT}
k0 LT0 LT1 0.9
k1 LT0 LT2 0.9
k2 LT1 LT2 0.9
.ENDS

** COMMENT OUT APPROPRIATE SECTIONS
** FOR DIFFERENT ANALYSES
** TRANSIENT ANALYSIS, Z, Q FACTOR
** TX. LINE BALUN FEED
T0 1 0 2 0 Z0={Z0} TD={TD}
XBALUN 2 3 4 BALUN
VTST0 3 5 DC 0.0 AC 0.0
VTST1 4 6 DC 0.0 AC 0.0
XPLDPA 5 6 PLNRDPA
VSIG 1 0 DC 0.001
+ SIN(0 {AMPL} {FREQ} 0 0 0)


** S PARAMETER MEASUREMENT COAX FEED
T0 1 0 2 0 Z0={Z0} TD={TD}
XPLDPA 2 0 PLNRDPA


** S PARAMETER MEASUREMENT TX. LINE BALUN FEED
T0 1 0 2 0 Z0={Z0} TD={TD}
XBALUN 2 3 4 BALUN
XPLDPA 3 4 PLNRDPA
VSIG 1 0 DC 0.001 AC 1.0


.OPTIONS METHOD=GEAR NOPAGE RELTOL-1m
** TRANSIENT ANALYSIS

.IC
.TRAN {TS} {TSTOP} {TSTRT} UIC
** LARGE SIGNAL CURRENT - VOLTAGE @ BOTH DIPOLE TERMINALS
.PRINT TRAN V(5) I(VTST0)
.PRINT TRAN V(6) I(VTST1)
```

```
.AC LIN {TOT} {FLLIM} {FHLIM}
** COAX FEED
.PRINT AC V(2)
** TX. LINE BALUN FEED
.PRINT AC V(3)
.PRINT AC V(4)
.END
```

The dipole antenna is a symmetrical device with two arms. The signal is fed into it using a combination of a transmission line and a balun (balanced/unbalanced). The signal is fed from the transmission line into the unbalanced side of the balun, and the balanced (two signals with the same amplitude and frequency, but 180 degree out of phase) outputs are fed into the two arms respectively of the dipole antenna. To determine the input impedance of each of the arms, **the RMS current flowing into a selected arm, and the voltage across that arm, is measured (using SPICE [2–5] transient analysis), and the ratio of the RMS voltage to the RMS current is the magnitude of the large signal input impedance of that arm**. The RMS voltage across and RMS current through one of the arms (labeled "A") are as below. The numbers in the first column are frequencies at which the measurements were made. The numbers in the last column are the radiation resistance, independent of the antenna. The planar dipole antenna is modeled as a series RLC circuit. Values of the input impedance of the selected arm of the planar dipole antenna as function of frequency are shown in Fig. 3.44. *As an antenna always needs to be impedance matched while in operation, knowledge of its input impedance is essential*:

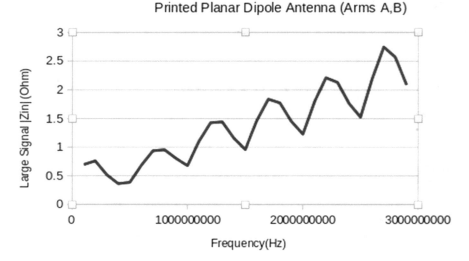

Fig. 3.44 Large signal input impedance magnitude of printed planar dipole antenna arms A, B

```
1.0E+8,4.434787e-02,6.358861e-02,12.325
2.0E+8,2.793168e-02,3.670152e-02,12.325
3.0E+8,2.324337e-02,4.509534e-02,12.325
4.0E+8,3.442942e-02,9.510466e-02,12.325
5.0E+8,5.719907e-01,1.487310e+00,12.325
6.0E+8,5.866776e-02,8.562300e-02,12.325
7.0E+8,4.953935e-02,5.288010e-02,12.325
8.0E+8,5.244396e-02,5.500701e-02,12.325
9.0E+9,7.903574e-02,9.831398e-02,12.325
1.0E+9,6.804493e-01,1.002203e+00,12.325
1.1E+9,9.773366e-02,8.852055e-02,12.325
1.2E+9,7.617230e-02,5.331882e-02,12.325
1.3E+9,7.979075e-02,5.524291e-02,12.325
1.4E+9,1.180537e-01,1.017110e-01,12.325
1.5E+9,6.869184e-01,7.168721e-01,12.325
1.6E+9,1.239733e-01,8.479713e-02,12.325
1.7E+9,9.854198e-02,5.359478e-02,12.326
1.8E+9,1.001611e-01,5.645073e-02,12.325
1.9E+9,1.533926e-01,1.057839e-01,12.325
2.0E+9,6.860777e-01,5.579074e-01,12.325
2.1E+9,1.581275e-01,8.827444e-02,12.325
2.2E+9,1.215835e-01,5.501569e-02,12.324
2.1E+9,1.581275e-01,8.827444e-02,12.325
2.2E+9,1.215835e-01,5.501569e-02,12.324
2.3E+9,1.256469e-01,5.901122e-02,12.325
2.4E+9,2.052609e-01,1.163017e-01,12.325
2.5E+9,6.959576e-01,4.554345e-01,12.325
2.6E+9,1.805488e-01,8.241627e-02,12.325
2.7E+9,1.464174e-01,5.331078e-02,12.325
2.8E+9,1.494776e-01,5.813129e-02,12.325
2.9E+9,2.329966e-01,1.117203e-01,12.325
3.0E+9,7.102064e-01,3.880070e-01,12.325
```

The RMS current through and RMS voltage across the printed planar dipole's other arm are listed as follows, and the large signal input impedance is shown in Fig. 3.44:

```
1.0E+8,4.434787e-02,6.358861e-02,12.325
2.0E+8,2.793168e-02,3.670152e-02,12.325
3.0E+8,2.324337e-02,4.509534e-02,12.235
4.0E+8,3.442942e-02,9.510467e-02,12.235
5.0E+8,5.719906e-01,1.487310e+00,12.325
6.0E+8,5.866776e-02,8.562300e-02,12.235
7.0E+8,4.953936e-02,5.288010e-02,12.325
8.0E+8,5.244395e-02,5.500701e-02,12.325
9.0E+8,7.903574e-02,9.831397e-02,12.325
1.0E+9,6.804493e-01,1.002203e+00,12.324
1.1E+9,9.773366e-02,8.852055e-02,12.325
1.2E+9,7.617230e-02,5.331881e-02,12.325
1.3E+9,7.979073e-02,5.524291e-02,12.325
```

```
1.4E+9,1.180537e-01,1.017110e-01,12.325
.5E+9,6.869185e-01,7.168722e-01,12.324
1.6E+9,1.239733e-01,8.479715e-02,12.325
1.7E+9,9.854197e-02,5.359478e-02,12.325
1.8E+9,1.001612e-01,5.645072e-02,12.325
1.9E+9,1.533926e-01,1.057839e-01,12.325
2.0E+9,6.860776e-01,5.579074e-01,12.325
2.1E+9,1.581275e-01,8.827444e-02,12.325
2.2E+9,1.215835e-01,5.501569e-02,12.325
2.3E+9,1.258750e-01,5.911834e-02,12.325
2.4E+9,2.053030e-01,1.163320e-01,12.325
2.5E+9,6.959576e-01,4.554345e-01,12.325
2.6E+9,1.805488e-01,8.241628e-02,12.325
2.7E+9,1.463182e-01,5.327467e-02,12.325
2.8E+9,1.494776e-01,5.813128e-02,12.325
2.9E+9,2.329966e-01,1.117203e-01,12.325
```

It is clear from Fig. 3.44 that the two arms of the printed dipole antenna are symmetrical, and the large signal input impedances of both the arms match. As the radiation resistance of the dipole antenna dominates over the frequency-dependent microstrip resistance, SPICE [2–5] small signal (AC) analysis can be performed—Fig. 3.45 shows the magnitude of the input reflection coefficient, obtained by SPICE [2–5] small signal analysis.

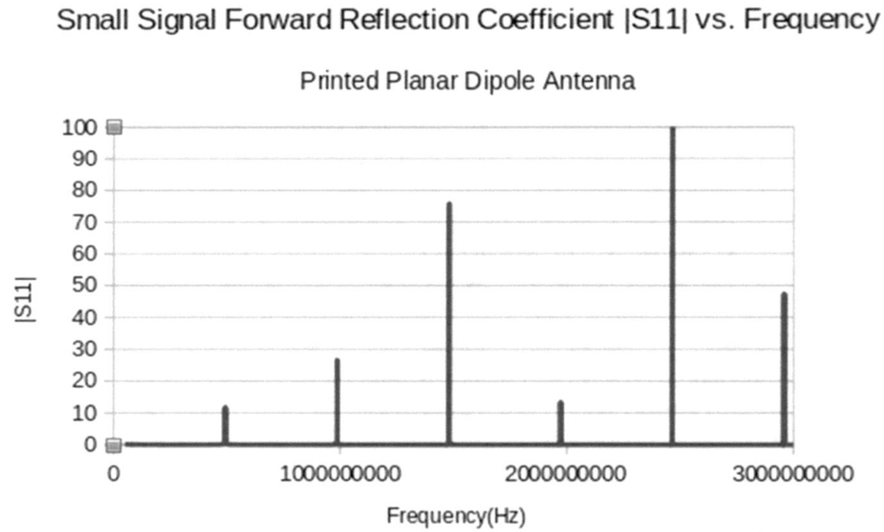

Fig. 3.45 Small signal forward reflection coefficient (dB) printed planar dipole antennas

3.13 Printed Planar Loop Antenna

The supplied C computer language [1] executable ***planarantlpms*** generates the SPICE [2–5] text input format netlist for the equivalent electrical circuit for a printed planar rectangular microstrip loop antenna. Although circular loop antennas are possible, rectangular|square loop antennas are easy to fabricate. Typing *./* **planarantlpms** at the Linux|MinGW shell command prompt generates the following help information:

```
./planarantlpms
incorrect|insufficient arguments
./planarantlpms b|B|c|C
<dielectric constant>
<frequency (MHz)>
<length mm>
<width mm>
<substrate thickness mm>
<trace thickness mm>
<trace width mm
sample batch|command line argument input
./planarantlpms b 9 434 40 25 1 0.035 1
```

Using the supplied sample batch|command line argument input at 100.0 MHz generates the following output:

```
 ./planarantlpms b 9 100 40 25 1 0.035 1
skin depth 6.201587e-06 m
loop inductance 1.026403e-07 H
radiation resistance 0.000384 Ohm
trace resistance 0.362704 Ohm
radiation and trace resistance total 0.363088 Ohm
capacitance 2.470376e-14 F
SPICE netlist planarantlpms.cir
```

The contents of the generated SPICE text input format netlist are:

```
PRINTED PLANAR LOOP ANTENNA

.PARAMS FREQ=1.000000e+08 FLLIM='0.1*FREQ' FHLIM='1.50*FREQ'
+ TS='1.0/(15.0*FREQ)' TSTOP='50.0/FREQ'
+ TSTRT='1.0/FREQ' AMPL=5.0 TOT=50000
.PARAMS C=2.470376e-14 L=1.026403e-07 RRAD=0.00038 RTR=0.36270
TD=1.0E-9 Z0=50.0

** PLANAR LOOP ANTENNA
.SUBCKT PLNLPANT 1
** 1 IN
C 2 0 {C}
L 2 3 {L}
```

```
RRAD 3 4 {RRAD}
RTRACE 4 0 {RTR}        .
 R0 1 2 1.0
.ENDS

** TRANSIENT ANALYSIS
T0 1 0 2 0 Z0={Z0} TD={TD}
XPLNLOOPANT 3 PLNLPANT
VTST0 2 3 DC 0.0 AC 0.0
VSIG 1 0 DC 0.001
+ SIN(0 {AMPL} {FREQ} 0 0 0)

.OPTIONS METHOD=GEAR NOPAGE RELTOL=1m
.IC
.TRAN {TS} {TSTOP} {TSTRT} UIC
.PRINT TRAN V(3) I(VTST0)
.END
```

SPICE [2–5] transient analysis is used to generate the large signal RMS current through and voltage across the printed planar rectangular loop antenna, obtained for each frequency in the selected frequency range of 100.0 MHz–2.0 GHz. The raw SPICE [2–5] transient analysis output is processed with the supplied C computer language [1] executable rmscalc, to extract the values for the RMS current and voltage, as listed in tabular format below:

```
1.0E+8,3.101642e+00,4.352611e-02,0.363088
2.0E+8,6.125753e+00,4.535222e-02,0.728994
3.0E+8,6.254376e+01,3.229108e-01,1.115369
4.0E+8,6.784291e+00,2.742308e-02,1.543997
5.0E+8,4.928568e+00,1.610113e-02,2.04713
6.0E+8,5.586564e+00,1.463373e-02,2.666246
7.0E+8,1.165884e+01,2.512856e-02,3.452307
8.0E+8,2.151804e+01,4.172261e-02,4.464416
9.0E+8,6.851238e+00,1.201114e-02,5.77251
1.0E+9,5.433816e+00,8.397458e-03,7.454666
1.1E+9,5.967212e+00,8.321821e-03,9.598445
1.2E+9,1.315346e+01,1.672440e-02,12.30627
1.3E+9,1.990049e+01,2.353627e-02,15.667209
1.4E+9,6.827215e+00,7.381974e-03,19.813405
1.5E+9,5.413487e+00,5.169066e-03,24.863647
1.6E+9,6.223756e+00,5.712368e-03,30.95851
1.7E+9,1.472656e+01,1.304308e-02,38.220073
1.8E+9,1.735815e+01,1.472914e-02,46.821206
1.9E+9,6.745771e+00,5.203704e-03,56.916279
```

The last column above represents the frequency-dependent total resistance of the rectangular printed planar loop antenna. The frequency-dependent input impedance and Q factor are shown in Figs. 3.46 and 3.47, respectively.

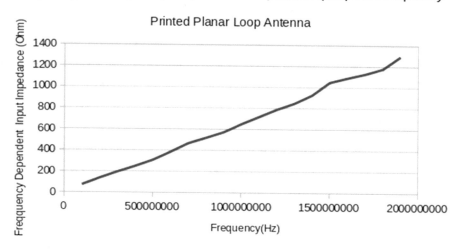

Fig. 3.46 Frequency-dependent input impedance printed planar rectangular loop antenna

Q Factor(Large Signal Model) vs. Frequency

Printed Planar Rectangular Loop Antenna

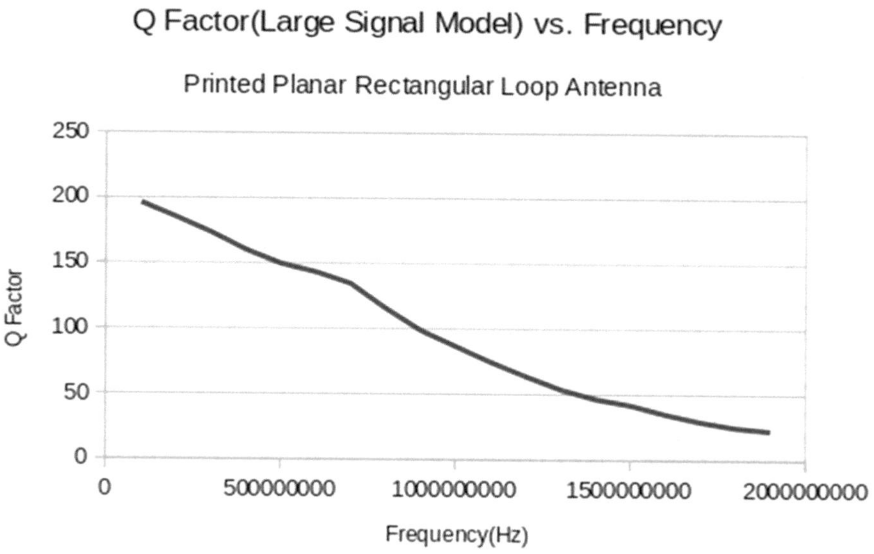

Fig. 3.47 Q factor printed planar rectangular loop antennas

3.14   Printed Planar Rectangular Patch Antenna: Quarter Wave Transmission Line Edge Connected Signal Feed—No Edge Shorting Pin

The supplied C computer language [1] executable ***planarantptchA*** generates the text SPICE [2–5] input format netlist for the equivalent electrical circuit (resonant RLC circuit) for the following four types of planar patch antenna:

- Quarter wave microstrip transmission line signal feed edge connected—no edge shorting pin/wall
- Impedance matched quarter microstrip wave transmission line inset signal feed—no edge shorting pin/wall
- Impedance matched quarter microstrip wave transmission line inset signal feed with edge shorting pin/wall
- Quarter wave microstrip transmission line signal feed edge connected with edge shorting pin/wall

Typing ./**planarantptchA** at the Linux|MinGW shell command prompt generates the following help information:

```
./planarantptchA
incorrect|insufficient input parameters
./planarantptchA b|B|c|C
<dielectric constant>
<operating frequency MHz>
<substrate thickness mm>
<patch antenna type 1|2|3|4>
patch antenna types
1 -> impedance matched inset microstrip signal feed - NO edge shorting
wall
2 -> edge quarter wave transmission line signal feed - NO edge shorting
wall
3 -> impedance inset microstrip signal feed - WITH edge shorting pin/wall
(inverted F antenna)
4 -> edge quarter wave transmission line signal feed - WITH edge shorting
pin/wall (inverted F antenna)
sample batch|command line input type 1
./planarantptchA b 4.5 500 0.1 1
sample batch|command line input type 2
./planarantptchA c 4.5 750 0.08 2
sample batch|command line input type 3
./planarantptchA b 4.5 650 0.1 3
sample batch|command line input type 4
./planarantptchA b 4.5 750 0.1 4
```

Using the supplied command line argument sample input for a quarter wave transmission line edge connected signal feed without edge shorting pin/wall, for a rectangular patch operating at 500 MHz antenna, generates:

```
./planarantptchA c 4.5 500 0.08 2
patch length 0.300 m
patch width 0.181 m
resonator capacitor 2.996315e-09 F
resonator inductor 3.384957e-11 H
resonator resistor 795.615 Ohm
SPICE netlist planarant.cir
```

The contents of the generated SPICE [2–5] netlist are:

```
EDGE CONNECTED QUARTER WAVE IMPEDANCE XFRMR MATCHED PATCH ANTENNA

.PARAMS C=2.996315e-09 L=3.384957e-11 R=795.615
.PARAMS TD=1.000000e-09 TDQWP=5.009691e-10 Z0QWP=199.451
.PARAMS FREQ=5.000000e+08 FLLIM='0.1*FREQ' FHLIM='1.5*FREQ'
+ Z0=50.0000 AMPL=1.5 RD=1.0
+ TS='1.0/(15.0*FREQ)' TSTOP='50.0/FREQ' TSTRT='1.0/FREQ' TOT=75000

** ANTENNA SUB-CIRCUIT
.SUBCKT PLNRPTCH 1
** 1 IN
C0 2 0 {C}
L0 2 0 {L}
R0 2 0 {R}
RD 1 2 {RD}
.ENDS

** TRANSIENT ANALYSIS Z Q Factor
T0 1 0 2 0 Z0={Z0} TD={TD}
T1 2 0 3 0 Z0={Z0QWP} TD={TDQWP}
XPLNR 4 PLNRPTCH
VTST0 3 4 DC 0.0 AC 0.0
VSIG 1 0 DC 0.001
+ SIN(0 {AMPL} {FREQ} 0 0 0)

** AC(SMALL SIGNAL) ANALYSIS
T0 1 0 2 0 Z0={Z0} TD={TD}
T1 2 0 3 0 Z0={Z0QWP} TD={TDQWP}
XPLNR 3 PLNRPTCH
VSIG 1 0 DC 0.001 AC 1.0

.OPTIONS METHOD=GEAR NOPAGE RELTOL=1m

** TRANSIENT ANALYSIS
.IC
.TRAN {TS} {TSTOP} {TSTRT} UIC
.PRINT TRAN V(4) I(VTST0)

** AC ANALYSIS
.AC LIN {TOT} {FLLIM} {FHLIM}
.PRINT AC V(3)
.END
```

SPICE [2–5] transient analysis is used to determine the RMS voltage across and current through the antenna over a selected range of frequencies of 100 MHz–2 GHz. These RMS current and voltage values can then be used to compute the frequency-dependent large signal input impedance The RMS current and voltage values are listed in tabular format below, where the last column is the resonator resistance:

```
1.0E+8,1.050042e-02,7.731826e-03,795.37
2.0E+8,3.171117e-02,2.201722e-02,795.432
3.0E+8,4.018014e-02,2.388121e-02,795.493
4.0E+8,2.078280e-02,9.141312e-03,795.552
5.0E+8,2.042037e-02,6.490942e-03,795.616
6.0E+8,3.117440e-02,9.039692e-03,795.677
7.0E+8,6.936328e-02,2.267786e-02,795.738
8.0E+8,1.012787e-01,2.422764e-02,795.799
9.0E+8,4.155796e-02,9.324592e-03,795.861
1.0E+9,3.266522e-02,6.336160e-03,795.992
1.1E+9,4.949613e-02,9.211403e-03,795.983
1.2E+9,1.082378e-01,2.357882e-02,796.045
1.3E+9,2.142574e-01,2.337731e-02,796.106
1.4E+9,6.573096e-02,9.011535e-03,796.167
1.5E+9,5.163250e-02,6.385929e-03,796.229
1.6E+9,7.216669e-02,9.167502e-03,796.290
1.7E+9,1.466786e-01,2.263865e-02,796.352
1.8E+9,3.745156e-01,2.165566e-02,796.418
1.9E+9,8.618019e-02,9.189687e-03,796.475
2.0E+9,7.358596e-02,6.224687e-03,796.546
2.1E+9,8.824949e-02,8.781452e-03,796.696
2.2E+9,1.817307e-01,2.145862e-02,796.659
2.3E+9,4.867647e-01,2.312188e-02,796.721
2.4E+9,1.186496e-01,8.840957e-03,796.782
```

The large signal frequency-dependent input impedance and small signal input reflection coefficient and forwardIinput return loss are in shown Figs. 3.48, 3.49 and 3.50.

3.15   Printed Planar Rectangular Patch Antenna: Quarter Wave Transmission Line Inset Signal Feed—Edge Shorting Pin (Inverted F Antenna)

The supplied C computer language [1] executable *planarantptchA* generates the text SPICE [2–5] input format netlist for the equivalent electrical circuit (resonant RLC circuit) for an impedance matched inset signal fed patch antenna with edge shorting pin wall. The SPICE netlist for such an antenna operating at 100 MHz is:

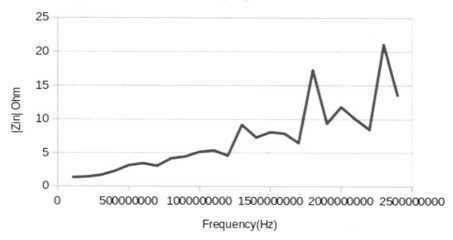

Fig. 3.48 Large signal input impedance rectangular patch antenna edge quarter wave transmission line feed—no edge shorting wall

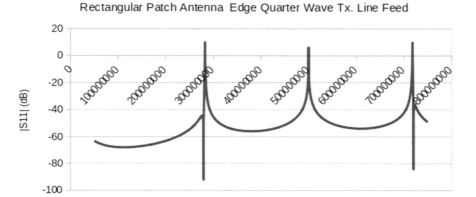

Fig. 3.49 Small signal forwardlinput reflection coefficient rectangular patch antenna edge quarter wave transmission line feed—no edge shorting wall

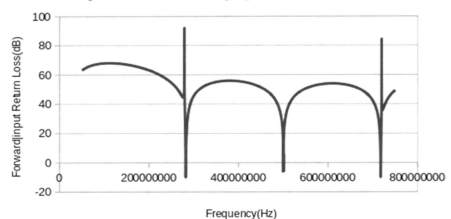

Fig. 3.50 Small signal forward|input return loss rectangular patch antenna edge quarter wave transmission line feed—no edge shorting wall

INSET QUARTER WAVE TX. LINE IMPEDANCE XFRMR MATCHED PATCH ANTENNA WITH
SHORTING WALL

```
.PARAMS C=9.656012e-07 L=2.625929e-12 R=49.449
.PARAMS FREQ=1.000000e+08 FLLIM='0.1*FREQ' FHLIM='1.5*FREQ'
+ Z0=50.000 Z0QWP=49.724 TD=1.000000e-09 TDQWP=2.501210e-09
LSW=1.866693e-06
+ AMPL=1.5 RD=1.0 TS='1.0/(15.0*FREQ)' TSTOP='50.0/FREQ' TSTRT='1.0/
FREQ' TOT=75000

** PATCH ANTENNA SUB-CIRCUIT
.SUBCKT PLNRPTCH 1
** 1 IN
C0 2 0 {C}
L0 2 0 {L}
LSW 2 3 {LSW}
R0 2 0 {R}
RD0 3 0 {RD}
RD1 1 2 {RD}
.ENDS

** TRANSIENT ANALYSIS Z Q
T0 1 0 2 0 Z0={Z0} TD={TD}
T1 2 0 3 0 Z0={Z0QWP} TD={TDQWP}
VTST0 3 4 DC 0.0 AC 0.0
XPLNRPTCH 4 PLNRPTCH
```

```
VSIG 1 0 DC 0.001
+ SIN(0 {AMPL} {FREQ} 0 0 0)

** S PARAMETER MEASUREMENT
T0 1 0 2 0 Z0={Z0} TD={TD}
T1 2 0 3 0 Z0={Z0QWP} TD={TDQWP}
XPLNRPTCH 3 PLNRPTCH
VSIG 1 0 DC 0.001 AC 1.0

.OPTIONS METHOD=GEAR NOPAGE RELTOL=1m

** TRANSIENT ANALYSIS
.IC
.TRAN {TS} {TSTOP} {TSTRT} UIC
.PRINT TRAN V(4) I(VTST0)

** S PARAMETER MEASUREMENT
.AC LIN {TOT} {FLLIM} {FHLIM}
.PRINT AC V(3)
.END
```

The large signal input impedance, frequency-dependent shorting pin/wall inductance, and small signal forward reflection coefficient are shown in Figs. 3.51, 3.52
and 3.53. In the above netlist, "LSW" is the shorting pin/wall inductance.

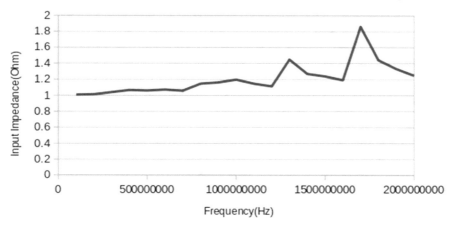

Fig. 3.51 Large signal input impedance matched inset feed rectangular patch antenna with shorting
pin/wall

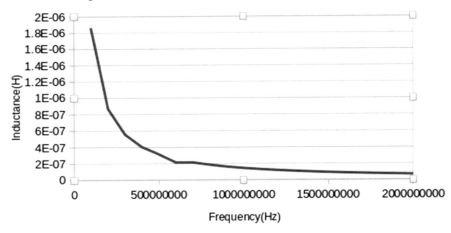

Fig. 3.52 Large signal impedance matched inset feed rectangular patch antenna with shorting pin/wall, inductance

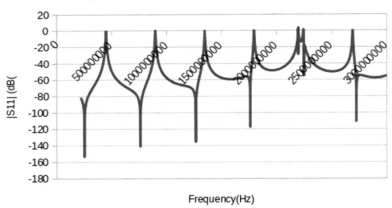

Fig. 3.53 Small signal forwardlinput reflection coefficient impedance matched inset feed rectangular patch antenna with shorting pin/wall

3.16 Printed Planar Rectangular Patch Antenna: Quarter Wave Transmission Line Inset Signal Feed—No Shorting Pin

The supplied C computer language [1] executable *planarantptchA* generates the text SPICE [2–5] input format netlist for the equivalent electrical circuit (resonant RLC circuit) for a for an inset signal fed patch antenna *without* edge shorting pin/wall. The SPICE netlist for such an antenna operating at 100 MHz is:

```
./planarantptchA b 4.5 100 0.1 1
patch length 1.501 m
patch width 0.905 m
ratio of signal input location to length 0.430
resonator capacitor 9.656012e-07 F
resonator inductor 2.625929e-12 H
resonator resistor 49.449 Ohm
SPICE netlist planarant.cir
```

where the contents of the SPICE [2–5] netlist are:

```
INSET QUARTER WAVE TX. LINE IMPEDANCE XFRMR MATCHED PATCH ANTENNA

.PARAMS C=9.656012e-07 L=2.625929e-12 R=49.449
.PARAMS FREQ=1.000000e+08 FLLIM='0.1*FREQ'
+ FHLIM='1.5*FREQ' Z0=50.000 Z0QWP=49.724 TD=1.000000e-09
TDQWP=2.501210e-09 AMPL=1.5 TS='1.0/(15.0*FREQ)' TSTOP='50.0/FREQ'
TSTRT='1.0/FREQ' TOT=75000

** PATCH ANTENNA SUB-CIRCUIT
.SUBCKT PLNRPTCH 1
** 1 IN
C0 2 0 {C}
L0 2 0 {L}
R0 2 0 {R}
RD 1 2 1.0
.ENDS

** TRANSIENT ANALYSIS
T0 1 0 2 0 Z0={Z0} TD={TD}
T1 2 0 3 0 Z0={Z0QWP} TD={TDQWP}
XPA 4 PLNRPTCH
VTST0 3 4 DC 0.0 AC 0.0
VSIG 1 0 DC 0.001
** S PARAMETER MEASUREMENT
T0 1 0 2 0 Z0={Z0} TD={TD}
T1 2 0 3 0 Z0={Z0QWP} TD={TDQWP}
XPA 3 PLNRPTCH
```

```
VSIG 1 0 DC 0.001 AC 1.0
.OPTIONS METHOD=GEAR NOPAGE RELTOL=1m
** TRANSIENT ANALYSIS
.IC
.TRAN {TS} {TSTOP} {TSTRT} UIC
.PRINT TRAN V(4) I(VTST0)

** S PARAMETER MEASUREMENT
.AC LIN {TOT} {FLLIM} {FHLIM}
.PRINT AC V(3)

.END
```

By design, the inset signal fed rectangular patch antenna is impedance matched, i.e., the attachment location of the quarter wave transmission line to the patch antenna is calculated so that impedance matching is ensured. Therefore, the large signal RMS voltage across and the current through the antenna at each selected frequency in the range 100 MHz 2 GHz are constant (with applicable tolerances). This is clear from the listing of the large signal RMS current through and voltage across the antenna as below:

```
1.0E+8,7.269924e-04,7.269886e-04
2.0E+8,6.493280e-04,6.493195e-04
3.0E+8,5.743289e-04,5.743162e-04
4.0E+8,5.422950e-04,5.422834e-04
5.0E+8,5.090940e-04,5.090792e-04
6.0E+8,4.732343e-04,4.732239e-04
7.0E+8,4.100314e-04,4.099862e-04
8.0E+8,3.973574e-04,3.973407e-04
9.0E+8,3.764611e-04,3.764205e-04
1.0E+9,3.669681e-04,3.669550e-04
1.1E+9,3.457384e-04,3.457020e-04
1.2E+9,3.430062e-04,3.429548e-04
1.3E+9,2.959336e-04,2.958957e-04
1.4E+9,2.918829e-04,2.918337e-04
1.5E+9,2.807754e-04,2.807192e-04
1.6E+9,2.729844e-04,2.729263e-04
1.7E+9,2.567226e-04,2.566132e-04
1.8E+9,2.414070e-04,2.412185e-04
1.9E+9,2.383627e-04,2.382323e-04
```

SPICE [2–5] small signal analysis is used to determine the small signal forward input reflection coefficient and input return loss as shown in Figs. 3.54 and 3.55, respectively.

Forward Reflection Coefficient

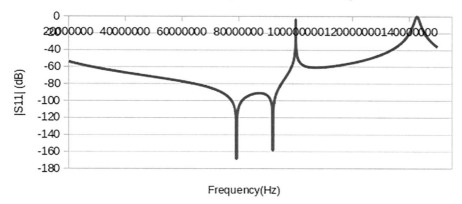

Fig. 3.54 Forward/input reflection coefficient inset signal fed impedance matched rectangular patch antenna

Input Return Loss

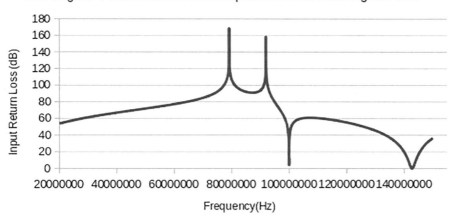

Fig. 3.55 Forward/input return loss inset signal fed impedance matched rectangular patch antenna

3.17　Printed Planar Rectangular Dual Parallel Gap Coupled Patch Antenna: Quarter Wave Transmission Line Inset Signal Feed—No Shorting Pin

The supplied C computer language [1] executable *planarantptch* generates the SPICE [2–5] text input format netlist for the equivalent electrical circuit of dual gap coupled rectangular patch antenna. The signal is fed into one patch (of the pair) through a quarter wave (wavelength of the receiveltransmission carrier wave) transmission line (for impedance matching), and this signal is then coupled into the neighbor via parasitic coupling between the two patches. Typing . */planarantptch* at the LinuxlMinGW shell command prompt generates the following help information:

```
./planarantptch
incorrect|insufficient input
./planarantptch b|B|c|C
<dielectric constant>
<frequency MHz>
<substrate thickness mm>
<trace thickness mm>
<signal input|shorting pin diameter mm>
<trace separation -- gap coupled parallel patches ONLY mm>
<shorting pin absent|present 1|2>
<single|dual parallel patch 0|1>
sample batch|command line input inset signal feed standalone patch
antenna NO shorting pin
./planarantptch B 9.8 500 1 0.01 0.0 0.0 1 0
sample batch|command line input inset signal feed standalone patch
antenna WITH shorting pin
./planarantptch B 9.8 500 1 0.01 0.025 0.0 2 0
sample batch|command line input inset signal feed gap coupled two patch
antenna NO shorting pin
./planarantptch c 9.8 500 1 0.01 0.0 0.01 1 1
sample batch|command line input inset signal feed gap coupled two patch
antenna WITH shorting pin
./planarantptch c 9.8 500 1 0.01 0.025 0.01 2 1
```

Using the supplied sample command line argument input for a dual patch gap coupled antenna with an operating frequency of 100.0 MHz generates the following output:

```
./planarantptch c 9.8 100 1 0.01 0.0 0.01 1 0
antenna length 1.47394 m
antenna width 0.64550 m
Eff. epsilon 1.034
Q factor 3050.118
capacitor 9.093546e-08 F
inductor 2.788351e-11 H
resistor 53.410 Ohm
```

```
shunt capacitor (gap coupled dual patch ONLY) 2.207904e-14 F
series capacitor 2.215912e-05 F
SPICE netlist planarant.cir
```

The contents of the generated SPICE [2–5] text input format netlist are listed below:

```
IMPEDANCE MATCHED GAP COUPLED TWO PATCH ANTENNA INSET SIGNAL FEED - NO
SHORTING PIN

.PARAMS C=9.093546e-08 L=2.788351e-11 FREQ=1.000000e+08
FHLIM='2.0*FREQ' FLLIM='0.1*FREQ'
+ TS='1.0/(15.0*FREQ)' TSTOP='50.0/FREQ' TSTRT='1.0/FREQ'
.PARAMS AMPL=2.5 CB=100.0 RP=53.4101 CGAP=2.215912e-05
CAPP=2.207904e-14
+ RD=1.0 TOT=75000
.PARAMS Z0=50.000 Z0QWP=51.677 TD=1.000000e-09 TDQWP=2.458921e-09
+ CPI=5.000000e-13

** GAP SEPARATED TWO PATCH ANTENNA SUB-CIRCUIT
.SUBCKT PLNRPTCHGC 1
** 1 IN
CP0 1 0 {C}
CP1 2 0 {C}
CGAP 1 2 {CGAP}
CAPP0 1 0 {CAPP}
CAPP1 2 0 {CAPP}
LP0 1 0 {L}
LP1 2 0 {L}
k0 LP0 LP1 0.95
RP0 1 0 {RP}
RP1 2 0 {RP}
.ENDS

** TRANSIENT ANALYSIS
T0 1 0 2 0 Z0={Z0} TD={TD}
T1 2 0 3 0 Z0={Z0QWP} TD={TDQWP}
VTST0 3 4 DC 0.0 AC 0.0
XPAGC 4 PLNRPTCHGC
VSIG 1 0 DC 0.001
+ SIN(0 {AMPL} {FREQ} 0 0 0)

VSIG 1 0 DC 0.001 AC {AMPL}

** AC (SMALL SIGNAL) S PARAMETER
T0 1 0 2 0 Z0={Z0} TD={TD}
T1 2 0 3 0 Z0={Z0QWP} TD={TDQWP}
XPAGC 3 PLNRPTCHGC
VSIG 1 0 DC 0.001 AC 1.0
```

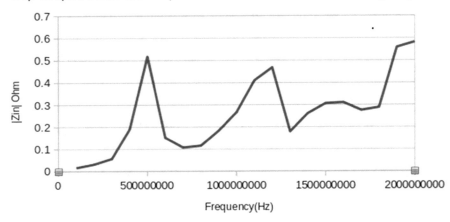

Fig. 3.56 Large signal input impedance gap coupled impedance matched dual patch antennas

```
.OPTIONS METHOD=GEAR NOPAGE RELTOL=1m
** TRANSIENT ANALYSIS
.IC
.TRAN {TS} {TSTOP} {TSTRT} UIC
.PRINT TRAN V(4) I(VTST0)
** AC(SMALL SIGNAL) S PARANETER
.AC LIN {TOT} {FLLIM} {FHLIM}
.PRINT AC V(3)
.END
```

SPICE [2–5] large signal transient analysis is used to determine the input impedance of the dual patch gap coupled antenna. However, as the antenna is already impedance matched, the input impedance is a constant (within applicable tolerances). The RMS voltage across and current through the antenna are listed below, in tabular format where the last column contains the Q (quality) factor at that frequency (100.0 MHz–2.0 GHz). Figures 3.56 and 3.57 show the large signal frequency-dependent input impedance and Q factors, respectively:

```
1.0E+8,8.697170e-04,5.354151e-02,3050.118
2.0E+8,4.053839e-03,1.276823e-01,1549.16
3.0E+8,1.149863e-02,2.016962e-01,1048.244
4.0E+8,1.192929e-02,6.246472e-02,797.368
5.0E+8,2.169791e-02,4.189335e-02,646.529
6.0E+8,7.722628e-03,5.059177e-02,545.724
7.0E+8,1.123653e-02,1.040587e-01,473.523
8.0E+8,3.669135e-02,3.155729e-01,419.208
9.0E+8,1.247337e-02,6.762229e-02,376.827
```

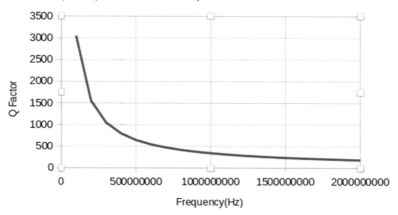

Fig. 3.57 Q factor large signal gap coupled impedance matched dual patch antennas

```
1.0E+9,1.231986e-02,4.626206e-02,342.805
1.1E+9,2.024458e-02,4.934607e-02,314.869
1.2E+9,4.330650e-02,9.245363e-02,291.501
1.3E+9,1.009855e-01,5.626329e-01,271.652
1.4E+9,1.835799e-02,7.033085e-02,254.571
1.5E+9,1.414078e-02,4.614898e-02,238.707
1.6E+9,1.430461e-02,4.605717e-02,226.647
1.7E+9,2.344399e-02,8.519542e-02,215.076
1.8E+9,1.271429e-01,4.411343e-01,204.747
1.9E+9,4.547265e-02,8.144283e-02,194.466
```

The small signal forwardǀinput reflection coefficient and return loss are shown in Figs. 3.58 and 3.59, respectively.

3.18   Printed Planar Rectangular Dual Parallel Gap Coupled Patch Antenna: Quarter Wave Transmission Line Inset Signal Feed—Shorting Pin

The supplied C computer language [1] executable *planarantptch* generates the SPICE [2–5] text input format netlist for the equivalent electrical circuit of dual gap coupled rectangular patch antenna. The signal is fed into one patch (of the pair) through a quarter wave (wavelength of the receiveǀtransmission carrier wave) transmission line (for impedance matching), and this signal is then coupled into the neighbor via capacitive coupling between the two patches. *The patch that is not*

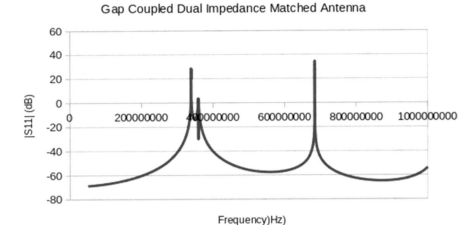

Fig. 3.58 Gap coupled impedance matched dual patch antenna small signal forward|input reflection coefficient

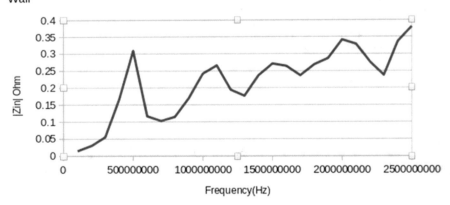

Fig. 3.59 Large signal input impedance gap coupled dual patch inset signal fed antenna with shorting pin on patch not connected to signal source

directly connected to the input signal quarter wave transmission line is grounded at
the end that is not capacitively coupled to the patch that is directly getting fed with
the signal via *the quarter wave transmission line.* Typing *. /planarantptch* at the
Linux|MinGW shell command prompt generates the following help information:

```
./planarantptch
incorrect|insufficient input
./planarantptch b|B|c|C
<dielectric constant>
<frequency MHz>
<substrate thickness mm>
<trace thickness mm>
<signal input|shorting pin diameter mm>
<trace separation -- gap coupled parallel patches ONLY mm>
<shorting pin absent|present 1|2>
<single|dual parallel patch 0|1>
sample batch|command line input inset signal feed standalone patch
antenna NO shorting pin
./planarantptch B 9.8 500 1 0.01 0.0 0.0 1 0
sample batch|command line input inset signal feed standalone patch
antenna WITH shorting pin
./planarantptch B 9.8 500 1 0.01 0.025 0.0 2 0
sample batch|command line input inset signal feed gap coupled two patch
antenna NO shorting pin
./planarantptch c 9.8 500 1 0.01 0.0 0.01 1 1
sample batch|command line input inset signal feed gap coupled two patch
antenna WITH shorting pin
./planarantptch c 9.8 500 1 0.01 0.025 0.01 2 1
```

Using the last supplied command line argument option generates the following
output for an operating frequency of 100.0 MHz:

```
./planarantptch B 9.8 100 1 0.01 0.025 0.01 2 1
antenna length 1.47394 m
antenna width 0.64550 m
Eff. epsilon 1.034
Q factor 3050.118
capacitor 9.093546e-08 F
inductor 2.788351e-11 H
resistor 53.410 Ohm
shunt capacitor(gap coupled dual patch ONLY) 2.207904e-14 F
series capacitor 2.215912e-05 F
SPICE netlist planarant.cir

IMPEDANCE MATCHED GAP COUPLED TWO PATCH ANTENNA INSET SIGNAL FEED -
SHORTING PIN

.PARAMS C=9.093546e-08 L=2.788351e-11 FREQ=1.000000e+08
FHLIM='2.0*FREQ' FLLIM='0.1%FREQ' TS='1.0/(15.0*FREQ)'
+ TSTOP='50.0/FREQ' TSTRT='1.0/FREQ'
.PARAMS AMPL=5.0 CB=100.0 RP=53.4101 CGAP=2.215912e-05
```

```
CAPP=2.207904e-14
+ RD=1.0 TOT=75000
.PARAMS Z0=50.000 Z0QWP=51.677 TD=1.000000e-09 TDQWP=2.458921e-09
+ LPF=3.017023e-15

** SUB-CIRCUIT FOR S PARANETERS
.SUBCKT PLNRDUALPTCHGSW 1
** 1 IN
CP0 1 0 {C
CP1 2 0 {C}
CGAP 1 2 {CGAP}
CAPP0 1 0 {CAPP}
CAPP1 2 0 {CAPP}
LP0 1 0 {L}
LP1 2 0 {L}
k0 LP0 LP1 0.95
LPF0 2 3 {LPF}
RP0 1 0 {RP}
RP1 2 0 {RP}
RD 3 0 {RD}
.ENDS

** TRANSIENT ANALYSIS
T0 1 0 2 0 Z0={Z0} TD={TD}
T1 2 0 3 0 Z0={Z0QWP} TD={TDQWP}
VTST0 3 4 DC 0.0 AC 0.0
XDPTCHSW 4 PLNRDUALPTCHGSW
VSIG 1 0 DC 0.001
+ SIN(0 {AMPL} {FREQ} 0 0 0)

** AC(SMALL SIGNAL) S PARAMETER
T0 1 0 2 0 Z0={Z0} TD={TD}
T1 2 0 3 0 Z0={Z0QWP} TD={TDQWP}
XDPTCHSW 3 PLNRDUALPTCHGSW
VSIG 1 0 DC 0.001 AC 1.0

.OPTIONS METHOD=GEAR NOPAGE RELTOL=1m
** TRANSIENT ANALYSIS
.IC
.TRAN {TS} {TSTOP} {TSTRT} UIC
.PRINT TRAN V(4) I(VTST0)

** SC(SMALL SIGNAL) S PARAMETER

.AC LIN {TOT} {FLLIM} {FHLIM}
** REFLECTION COEFFICIENT, VSWR
.PRINT AC V(3)

.END
```

The large signal RMS values for the current flowing through and voltage across the gap coupled dual patch inset signal feed (impedance matched) and shorting pin (on coupled patch) antenna that were obtained using SPICE [2–5] transient analysis are listed below. The selected frequency range is 100 MHz–2.5 GHz. For each selected frequency in this range, the RMS voltage across, the current through, and Q factor are measured. The raw SPICE [2–5] transient analysis output, for each frequency, is processed with the supplied C computer language [1] executable **rmscalc** to determine the corresponding RMS current and voltage from that frequency:

```
1.0E+8,1.610895e-03,1.089475e-01,3058.511
2.0E+8,7.855359e-03,2.601827e-01,1549.16
3.0E+8,2.261009e-02,4.082663e-01,1048.244
4.0E+8,2.014376e-02,1.210687e-01,797.368
5.0E+8,2.518687e-02,8.152586e-02,646.529
6.0E+8,1.185684e-02,1.012678e-01,545.724
7.0E+8,2.204316e-02,2.139463e-01,473.523
8.0E+8,7.027611e-02,6.100015e-01,419.208
9.0E+8,2.312421e-02,1.362887e-01,376.827
1.0E+9,2.181253e-02,9.007538e-02,342.805
1.1E+9,2.588279e-02,9.724312e-02,314.868
1.2E+9,3.694125e-02,1.902504e-01,291.501
1.3E+9,1.962581e-01,1.111846e+00,271.652
1.4E+9,3.275985e-02,1.382475e-01,254.571
1.5E+9,2.537867e-02,9.356621e-02,239.707
1.6E+9,2.535222e-02,9.594783e-02,226.647
1.7E+9,4.061421e-02,1.719241e-01,215.076
1.8E+9,2.392093e-01,8.903852e-01,204.747
1.9E+9,4.628841e-02,1.613090e-01,195.466
2.0E+9,3.270201e-02,9.593232e-02,187.077
2.1E+9,2.833104e-02,8.631516e-02,179.455
2.2E+9,4.258031e-02,1.542202e-01,172.496
2.3E+9,2.126475e-01,8.974082e-01,166.114
2.4E+9,5.944619e-02,1.770206e-01,150.239
2.5E+9,3.828354e-02,1.008588e-01,154.811
```

The large signal frequency-dependent input impedance, Q factor, and small signal forward reflection coefficient and return loss are shown in Figs. 3.59, 3.60, 3.61, and 3.62, respectively.

Q Factor vs. Frequency

Gap Coupled Dual Patch Impedance Matched Antenna Shorting Pin|
Wall

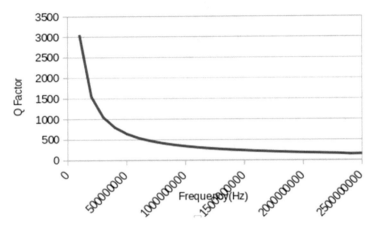

Fig. 3.60 Large signal Q factor gap coupled dual patch inset signal fed antenna with shorting pin on patch not connected to signal source

Forward|Input Reflection Coefficient vs, Frequency

Gap Coupled Dual Patch Inset Feed Antenna Shorting Pin|Wall

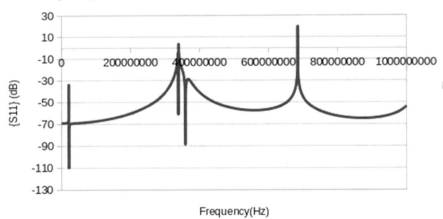

Fig. 3.61 Small signal forward reflection coefficient gap coupled dual patch impedance matched antenna

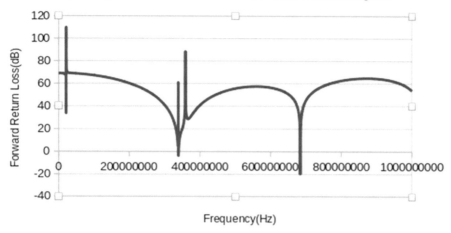

Fig. 3.62 Small signal forward return loss gap coupled dual patch impedance matched antenna

References

1. The all time classic C computer language book by its creators https://www.amazon.in/Programming-Language-Prentice-Hall-Software/dp/0131103628
2. Ngspice user manual ttp://ngspice.sourceforge.net/docs/ngspice-manual.pdf
3. Pspice user manual https://resources.pcb.cadence.com/i/1180526-pspice-user-guide/25?
4. Hspice user manual https://picture.iczhiku.com/resource/eetop/SHitrwIJSQFzExcV.pdf
5. TINA Spice user manual https://www.ti.com/lit/ug/sbou052a/sbou052a.pdf?ts=16366102333 62&ref_url=https%253A%252F%252Fwww.google.com%252
6. https://onlinelibrary.wiley.com/doi/full/10.1002/jnm.2272

Appendix A: Ready-to-Use ANSI C Computer Language Executables for Both Linux and Windows Operating Systems

Analysis and design examples in Chap. 3 demonstrate how to use these executables with appropriate command line arguments. Developed on a Fedora 32 operating system machine, these have been ported to Ubuntu 18.04 LTS, and Windows 11 operating systems with MinGW firmware installed.

- **planarind**
- **planarindd**
- **planarindgh**
- **planarindls1**
- **planarindmw**
- **planarindxfrmr**
- **planarantdp**
- **planarantlpms**
- **planarantptchA**
- **planarantptch**
- **sparameasdb**
- **rmscalc**
- **qfactmeas**

© The Editor(s) (if applicable) and The Author(s), under exclusive license to
Springer Nature Switzerland AG 2023
A. Banerjee, *Planar Spiral Inductors, Planar Antennas and Embedded Planar Transformers*, https://doi.org/10.1007/978-3-031-08778-3

Appendix B: How to Download and Install the MinGW Firmware on Any Windows Operating System Computer

MinGW (Minimalist GNU for Windows) is a freely downloadable firmware package that enables the user to use the reliable, robust, and widely used *gcc* compiler suite on any Windows operating system machine. The following Web sites (URLs) contain step-by-step instructions with screenshots on how to install MinGW on any Windows operating system machine.

https://genome.sph.umich.edu/wiki/Installing_MinGW_%26_MSYS_on_Windows

https://www.ics.uci.edu/~pattis/common/handouts/mingweclipse/mingw.html

https://www.rose-hulman.edu/class/csse/resources/MinGW/installation.htm

https://sourceforge.net/projects/mingw-w64/

http://www.codebind.com/cprogramming/install-mingw-windows-10-gcc/

© The Editor(s) (if applicable) and The Author(s), under exclusive license to
Springer Nature Switzerland AG 2023
A. Banerjee, *Planar Spiral Inductors, Planar Antennas and Embedded Planar Transformers*, https://doi.org/10.1007/978-3-031-08778-3

Index

© The Editor(s) (if applicable) and The Author(s), under exclusive license to
Springer Nature Switzerland AG 2023
A. Banerjee, *Planar Spiral Inductors, Planar Antennas and Embedded Planar
Transformers*, https://doi.org/10.1007/978-3-031-08778-3

Printed in the United States
by Baker & Taylor Publisher Services